INTERNATIONAL CENTRE FOR MECHANICAL SCIENCES

COURSES AND LECTURES - No. 236

M. ESSLINGER - B. GEIER
INSTITUT FÜR FLUGZEUGBAU
BRAUNSCHWEIG

POSTBUCKLING BEHAVIOR OF STRUCTURES

SPRINGER-VERLAG WIEN GMBH

ISBN 978-3-211-81369-0   ISBN 978-3-7091-2932-6 (eBook)

DOI 10.1007/978-3-7091-2932-6

1. INTRODUCTION

Aeroplanes have to be extremely light, otherwise they will
not be able to fly. Hence, aircraft engineers must deal
with light-weight structures which are usually thin-walled
and liable to buckling if subjected to compressive load.
Thus, in aircraft engineering structural stability plays
an important rôle.

The structural ideas and calculation methods of the aircraft
engineers are utilized also by civil engineers and are sup-
plemented by them, with respect to their specific applications, whenever an economic advantage can be achieved. In
the course of this trend to improved economy the maximum
capacity load, i.e. the load at which the structure actually fails, is introduced into the strength calculations,
rather than the admissible stress. With respect to the maximum capacity load a certain safety margin must be observed.

For statically determinate structures loaded in tension the
maximum capacity load is attained as soon as the yield limit
has been reached. For statically indeterminate structures
the load at which the yield limit is attained locally, may
be exceeded, especially under bending stresses.

For structures loaded in compression a proof of sufficient
stability has to be performed. The buckling load is not
always the maximum capacity load. There are indeed struc-

tures, whose load-carrying capacity is not exhaustet with
the appearance of visible buckles. Whether the load carried
by the structure can be augmented after buckling or not,
depends on the postbuckling behavior.

An example for safe postbuckling behavior, based on your
personal empirical data: When a car-driver finds a visible
buckle in a fender of his car, he will be sad at the loss
of glamour. But he knows that this buckle has not yet de-
teriorated the safety of his car.

For an expert on shell structures a very unpleasant situation comes up if he is led to a structure with a visible
buckle and has to respond to the anxiously anticipated question: "Can you see the buckle up there? Will it do any harm?"

The postbuckling behavior not only is of interest for the
load carrying capacity after buckling, but it also influ-
ences the magnitude of the actual buckling load. In order
to explain these relationships, we are going to compare the
buckling and postbuckling behavior of three extremely different structures: Column, plate and cylinder.

Fig.1.1 shows, on the left hand side, the load shortening
curve of a column. In the prebuckling region the load increases in proportion to the shortening. When the buckling
load has been attained the column deflects laterally. The
deflecting as well as the restoring forces to the first approximation are linear functions of the deflection w . For
this reason the column subjected to the buckling load is in
equilibrium at any deflection w . The postbuckling curve
is a horizontal line.

We come to the flat plate, Fig.1.2. In the prebuckling region the load increases in proportion to the shortening.

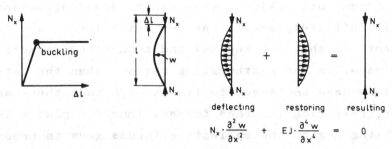

$$N_x \cdot \frac{\partial^2 w}{\partial x^2} + EJ \cdot \frac{\partial^4 w}{\partial x^4} = 0$$

Fig. 1.1 Buckling and postbuckling behavior of a perfect column.

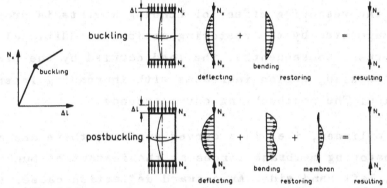

Fig. 1.2 Buckling and postbuckling behavior of a perfect plate.

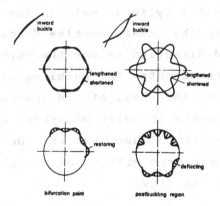

Fig. 1.3 Membrane forces in a buckled circular cylinder.

At the instant of buckling, when the lateral deflections
are still infinitely small, the restoring forces are bending moments in the longitudinal and transverse direction,
nothing more. In the postbuckling region, when the lateral
deflections have increased to finite magnitude there arise,
moreover, restoring membrane forces. Thus for plates in the
postbuckling region the deflecting forces grow in proportion
to the lateral deflection w as with the column, whereas
the restoring forces increase faster than proportionally
because the restoring effect of bending moments is progres-
sively augmented by the restoring action of additional membrane forces. Consequently, the load carried by the plate in
the postbuckling region increases with increasing lateral
deflection. The postbuckling curve ascends.

For the cylinder, i.e. in a curved surface, there are already restoring membrane forces at the instant of buckling,
Fig.1.3, left hand side. An outward deflection causes tensile forces, an inward deflection causes compressive forces.
These membrane forces in combination with the curvature of
the cylinder act in the restoring sense. Tension in the outward buckles forces the cylinder wall in the inward direction; compression in the inward buckles forces the cylinder
wall in the outward direction (since the curvature of the
undeformed shell predominates the buckling deformation).
From the fact that in the shell at the instant of buckling
not only bending moments but also membrane forces act in
the restoring sense, it follows that the buckling loads of
shells are high as compared with those of columns of plates
with equal flexural rigidity.

In the postbuckling region the radial deflections become so
large that they no longer may be neglected in comparison to
the curvature of the undeformed shell, Fig.1.3, right hand

side. When the inward postbuckling deformation predominates
the initial shell curvature, the membrane compressive forces
in the inward buckles push the shell wall no longer in the
outward but in the inward direction. Now the membrane forces
produced by the buckling process yield radial loads in the
inward direction over the whole circumference. The effect
of these swelling radial loads is no longer restoring, but
on the contrary is detrimental like external pressure. From
the fact that for thin-walled shells in the postbuckling
region the restoring action of the membrane forces decreases,
vanishes and finally is reversed, it follows that the loads,
carried by the cylinder in the postbuckling region, are lower than the buckling load.

Fig.1.4 shows the load-shortening curve of a perfect cylinder, loaded by controlled shortening. In the prebuckling region the load grows in proportion to the shortening. As soon
as the buckling load, or strictly speaking, the buckling shortening is attained, the load carried by the cylinder suddenly
drops with unchanged shortening. The cylinder wall assumes
a new stable equilibrium state at an axial load far below
the buckling load. During the transition from the unbuckled
to the stable buckled equilibrium state a number of unstable
deformation states are passed.

We turn to imperfect structures, Fig.1.5. The imperfect column approaches the buckling load only when the deformation
becomes very large. The imperfect plate attains and surpasses
the buckling load of the perfect plate in the region of finite shortening. The imperfect cylinder does not attain the
buckling load of the perfect cylinder, since the initial imperfections locally reduce the restoring effect of the membrane forces.

The fact that thin-walled shells are imperfection-sensitive

8

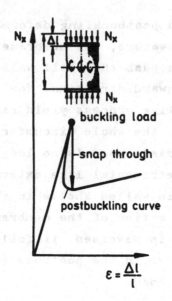

Fig.1.4 Buckling and postbuckling behavior of a perfect cylinder.

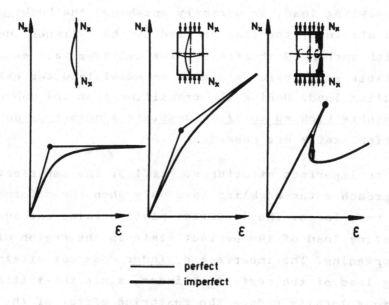

perfect
imperfect

Fig.1.5 Buckling and postbuckling behavior of imperfect structures.

due to postbuckling equilibrium states with small finite
deflections at loads far below the buckling load, influe-
ences the practical engineering work with regard to three
aspects:

- In proportioning thin-walled compressed shells the
  theoretical buckling load of the perfect shell must
  not be considered as the load carrying capacity. Instead, a reduced value has to be introduced. The reduction factor depends on the shape of the shell and
  on the type of loading. The actual buckling load in
  the most unfavourable case can be as low as 20 % of
  the theoretical buckling load of the perfect shell.

- In choosing the safety factors one has to regard the
  level of risk. This means that to an axially loaded
  circular cylindrical shell, the load carrying capacity
  of which drops at buckling, a higher safety factor has
  to be prescribed, than to a flat plate, which in the
  postbuckling region may carry several times the buckling load.

- Optimization studies must not be performed by comparing
  theoretical buckling loads of perfect shells, but rather
  with the imperfection sensitivity taken into account.
  This can appropriately be done by using a suitably defined postbuckling load as an index value.

These aspects will be elaborated in detail in the following.
The paper is restricted to two-dimensional structures. There
is one chapter for plates and folded plate structures, one
for thin-walled cylinders and one for spherical shells. In
these three chapters experimental and theoretical results
will be discussed. The differential equations and solution
procedures will be treated in the final chapter.

2. PLATES AND FOLDES STRUCTURES

2.1 Rectangular plates

2.1.1 Effective width

Fig.2.1 presents panels and folded structures, subjected to
compression forces in one direction only. In such structures
the compressive stresses are not uniformly distributed, but
are lower at the free edges or in the center region of the
panels than at the corners. For the calculation of stress
and deformation in panels under compression, engineers all
over the world use the concept of effective width, that
means the width of a partial panel, which transmits the
same force as the whole panel, but has constant stress; in
fact the same as at the stiffened panel edges. Fig.2.2 shows
the effective width for an example problem. For thin-walled
highly loaded panels it can drop to 5 % or less of the actual width.

There are two reasons for the non-uniform stress distribution
in the panels and, consequently, two fundamentally different
methods for calculating the effective width:

● In the case of the reinforced sheet, Fig.2.1a, the compressive forces are applied locally. Their diffusion
  into the sheet is related to deformation energy. The
  wider the zone over which the stress in the panels is
  distributed, the lower the strain energy of the compressive longitudinal stresses themselves, and the

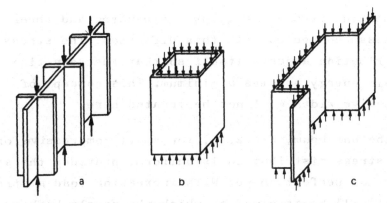

Fig.2.1 Schematic presentation of structures composed of flat plates.

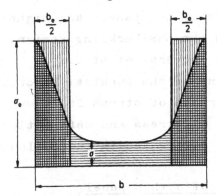

Fig.2.2 Schematic presentation of effective width.

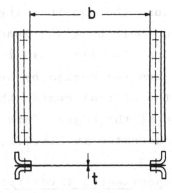

Fig.2.3 Flat panel with edge stiffeners.

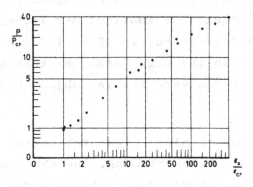

Fig.2.4 Experimental postbuckling loads.

higher the strain energy of the bending and shear
stresses connected with this diffusion. The stress
distribution adjusts itself so that the resulting
strain energy becomes a minimum. This concept of
effective width will not be treated here.

● In the box beam, Fig.2.1b, for small compressive forces
the stress distribution is uniform, provided the structure has perfect shape. With increasing load a stress
level will be attained at which the panels buckle.
Notwithstanding, the load can be increased further,
because the postbuckling loads of the panels are higher
than their buckling loads. In the postbuckling region
the panels will avoid further increase of stress in
the central region by deepening of the buckles; hence,
in the central region the increase of stress is lower
than at the edges. This state of stress and deformation,
demonstrated in Fig.2.2, will be treated in the following.

2.1.2 Experimental investigations of a thin panel

We begin with test results obtained for very thin panels
with rigid edge beams, Fig.2.3, by H. Wagner [1] and his
staff in the Flugtechnisches Institut der Technischen Hochschule, Berlin-Charlottenburg, in 1936. The panel was compressed in parallel to the edge beams, so that its whole
width and the edge beams were always subjected to the same
shortening ε_x .

Fig.2.4 shows in dimensionless presentation the postbuckling
loads as function of the shortening. The abscissa presents
the ratio $\varepsilon_x / \varepsilon_{cr}$, where ε_{cr} is the theoretical buckling
shortening of the perfect panel and ε_x the actual short-
ening of the panel and the edge beams. The tests begin at

the buckling shortening $\epsilon_x / \epsilon_{cr} = 1$ and run deeply into
the postbuckling region, that means up to a shortening ϵ_x
which is 400 times as large as the buckling shortening ϵ_{cr}.
The ordinate shows P/P_{cr}, where P_{cr} is the theoretical
buckling load of the perfect panel and P the postbuckling
load of the panel, that is the load carried by the buckled
panel, exclusive of the edge beams.

At the buckling shortening $\epsilon_x / \epsilon_{cr} = 1$, the experimental
load, carried by the panel, is somewhat lower than the theoretical buckling load, because the panel was weakened by
initial imperfections and initial stresses. If the shortening is 400 times as large as the buckling shortening,
then the postbuckling load of the panel is 40 times as large
as the buckling load. The fact that the postbuckling load
increases more slowly than the postbuckling shortening is
due to the phenomenon that the center region of the panel
avoids more and more an increase of stress by deepening of
the buckling amplitudes; the effective width becomes smaller
and smaller with increasing load.

Fig.2.5 is a schematic presentation of the variation of the
compressive membrane stress across the panel width at different levels of postbuckling shortening. This diagram is
valid for perfect panels only. The ordinate gives the ratio

$$\frac{\sigma_x}{\sigma_e} = \frac{\text{local stress in the panel}}{\text{stress in the edge beam}} \quad .$$

Curve parameter is the ratio

$$\frac{\epsilon_x}{\epsilon_{cr}} = \frac{\text{shortening of the whole structure}}{\text{buckling shortening of the panel}} \quad .$$

For the shortenings $\epsilon_x / \epsilon_{cr} > 50$ the central region of

14

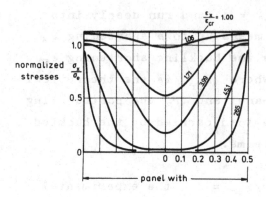

normalized
stresses $\frac{\sigma_x}{\sigma_e}$

panel with

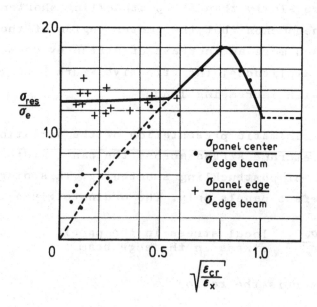

**Fig.2.5 Schematic presen-
tation of the variation of
compressive stress with width.**

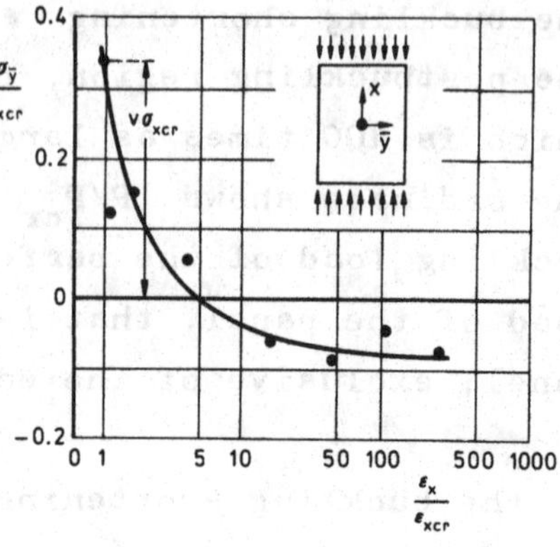

**Fig.2.6 Membrane stresses
in $\bar{y}$-direction.**

$\frac{\sigma_{res}}{\sigma_e}$

$\frac{\sigma_{panel\ center}}{\sigma_{edge\ beam}}$

$\frac{\sigma_{panel\ edge}}{\sigma_{edge\ beam}}$

$\sqrt{\dfrac{\varepsilon_{cr}}{\varepsilon_x}}$

Fig.2.7 Resulting stresses in the buckled panel.

the panel hardly participates in an augmentation of the compressive force. It seems probable, that with further increase of ε_x the effective width becomes independent of the actual width. At buckling $\varepsilon_x / \varepsilon_{cr} = 1$, the theoretical stress in the panel is higher than the stress in the edge beams. This difference is due to the fact that the distance of the edge beams is assumed to be kept constant, so that a two dimensional state of stress without transverse strain occurs in the panel. As against Young's modulus in the edge beams being E, the modulus in the panels is $\frac{E}{1-\nu^2}$.

In the tests, the distance of the edge beams was indeed fixed, so that prior to buckling due to Poisson's ratio compressive stresses, transverse to the external loading acted in the panel. Fig.2.6 shows these transverse stresses as function of the dimensionless postbuckling shortening. When the panel buckles, the compressive transverse stresses are reduced. At five times the buckling shortening, they are exactly zero. When the panel is shortened further, that is in the advanced postbuckling region, restoring tensile stresses occur. Please, remember the introduction, where we pointed out that supporting membrane forces occur at the plate in the postbuckling region.

Now the question arises as to what values the stresses in the panel increase in the postbuckling domain. This question is answered by Fig.2.7. The abscissa shows the ratio

$$\sqrt{\frac{\varepsilon_{cr}}{\varepsilon_x}} = \sqrt{\frac{\text{buckling shortening of the panel}}{\text{shortening of the whole structure}}} \quad .$$

The test results begin from the right hand side at the buckling shortening $\sqrt{\frac{\varepsilon_{cr}}{\varepsilon_x}} = 1$ and run on to the left hand side up to a shortening, which is 400 times as large as the buckling shortening. The ordinate presents the ratio

$$\frac{\sigma_{res}}{\sigma_e} = \frac{\text{membrane plus bending stress in the panel}}{\text{compressive stress in the edge beams}}$$

On the right hand side, prior to buckling till buckling, the stress in the perfect panel would be about 10 % higher than the stress in the edge beams, since in the panel, a two dimensional state of stress without transverse strain occurs. But this is theory; test points are not plotted in this part of the diagram. When the panel buckles, the stress in the center region at first increases more rapidly than the stress in the edge beams, since the bending stresses of the postbuckling deformation are added to the membrane compressive stresses. This increase of stress in the center region of the panel is counteracted by the inclination of the buckled panel to avoid further augmentation of the compressive membrane stress by deepening the buckles. From these opposite tendencies a maximum of the normalized panel stress results at $\sqrt{\frac{\varepsilon_{cr}}{\varepsilon_x}} = 0.80$, i.e. at 1.5 times the buckling shortening. At this maximum, the resulting stress in the panel is 1.8 times as high as the stress in the edge beams. With further shortening the normalized stress decreases, since in the middle region the panel more and more successfully avoids a further increase of load. In spite of this decrease of the normalized stress the absolute panel stress increases further.

With $\sqrt{\frac{\varepsilon_{cr}}{\varepsilon_x}} = 0.55$, i.e. when the actual shortening amounts to 3.0 times the buckling shortening the panel stress in the vicinity of the edge beams comes up to the stress in the center region. Here the normalized panel stresses are 1.25 and this value remains constant at the panel edges with inbreasing shortening.

In the center region the normalized stress decreases further
with increasing shortening, since the absolute stress increases more slowly in the middle of the panel than in the
edge beams.

The effective width, Fig.2.2, is defined by the ratio

$$b_e \cdot t = \frac{\text{load carried by the panel}}{\text{stress in the edge beams}} \; .$$

Fig.2.8 gives a dimensionless representation of the effec-
tive width as a function of shortening. At buckling $\epsilon_x / \epsilon_{cr} = 1$
the theoretical effective width of the perfect panel would
be $b_e/b > 1$ since in the panel a higher compressive stress
prevails than in the edge beams on account of Poisson's ratio. But in the actual panel, the effective width is $b_e/b < 1$
already at the instant of buckling, since the load carried
by the panel is reduced by initial imperfections and initial
stresses. The effective width decreases with increasing
shortening and amounts to 4 % of the actual panel width at
$\epsilon_x / \epsilon_{cr} = 400$.

The tests were performed with edge beams in fixed distance.
Complementary tests with freely shiftable edge beams have
shown that in this case the effective widths are only
slightly smaller. The difference is almost within the
scatter of test results.

The tests were performed at panels with clamped edges. Due
to theoretical considerations [2] it seems probable that the
test results can also be applied to panels with simply supported edges. For this transfer one must proceed from the
dimensionless plot of Fig.2.8, and take into account that
the buckling shortening ϵ_{cr} of the panel, used on the abscissa strongly depends on the boundary conditions. From

18

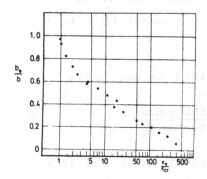

Fig.2.8 Experimental
effective width for stress
calculation.

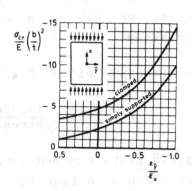

Fig.2.9 Theoretical buck-
ling stress for different
edge conditions.

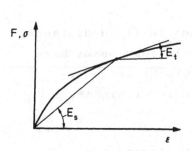

Fig.2.10 Schematic pres-
entation of a load-short-
ening curve.

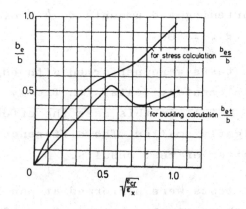

Fig.2.11 Experimental effective width for stress and
buckling calculation.

Fig.2.9 it can be seen, that for vanishing transverse strain, the buckling load of the clamped panel is twice as high as that of the simply supported one. For the buckling short-ening ε_{cr} the same relation is valid.

The above reflections concerning the effective width are sufficient for the calculations of the s t r e s s e s . But they need completion, when the s t a b i l i t y of folded structures, Figs. 2.1b and c , is considered. This shall be explained with the aid of Fig.2.10. Here you see a schematic presentation of a load shortening curve. If this curve is valid for a strut, the stress of which, has exceeded the limit of elasticity, it is customary to dis-tinguish between the secant modulus E_s , which indicates the ratio of load and shortening, and the tangent modulus E_t , which is the criterion for the stress variation due to a small variation of shortening. Correspondingly, in the case of a panel in the postbuckling region, it must be distin-guished between the effective width for the calculation of s t r e s s

$$b_{es} = \frac{1}{E \cdot t} \cdot \frac{F}{\varepsilon} ,$$

and the effective width for the calculation of s t a b i l - i t y

$$b_{et} = \frac{1}{E \cdot t} \cdot \frac{dF}{d\varepsilon} .$$

The effective width b_{et} for the calculation of stability is always smaller than the effective width b_{es} for the calculation of stress. The two effective widths determined by the tests are given in Fig.2.11. At the moment of buck-ling the ratio is smallest

$$\frac{b_{et}}{b_{es}} = 0.5 .$$

With increasing shortening the two values approach each other.

The load carrying capacity of the panel in the postbuckling
region, is not exhausted before the membrane stresses attain
the yield stress, that is, the buckled panels do not fail,
until the supporting edge beams are stressed up to about
90 % of the limit of plasticity.

2.1.3 Theoretical investigations of a perfect panel

Theoretical postbuckling investigations are rather complex,
since the postbuckling deformations are finitely large and
one has to deal with a nonlinear plate theory. Generally the
states of equilibrium and deformation in the postbuckling
region are represented by series expansions or with the aid
of finite elements. Series expansions are preferred, since
they allow definite assumptions on the shape of the postbuckling pattern.

It would be possible to calculate the effective width of
perfect panels with great accuracy, using computers. But
this effort would not be reasonable, since perfect panels
do not exist and imperfect panels can be calculated with
sufficient accuracy using rough approximations.

In the following, a theoretical paper published by Marguerre
[3] in 1937 will be discussed. The calculation is based on
a rectangular plate, which is simply supported at the four
edges on rigid edge beams. The longitudinal edge beams can
translate freely.

In a first approximation it has been assumed, that the shape
of the buckling pattern

$$w = a \cdot \cos \frac{\pi x}{l} \cdot \cos \frac{\pi y}{b}$$

remains preserved, when the buckling load is exceeded. The

calculation with this simple postbuckling pattern yields
that the load shortening curve after buckling ascends only
half as steeply as prior to buckling, that is the effective
width b_{et} , for the calculation of stability, after buckling is half as large as prior to buckling, Fig.2.12. The
effective width for the calculation of stress, does not
change at the moment of buckling. Therefore in the prebuckling region

$$b_{et} = b_{es} \text{ ,}$$

and in the initial postbuckling region

$$b_{et} = 0.5 \; b_{es} \text{ .}$$

This result found by means of an approximate analysis corresponds well to the test results presented in Fig.2.11 .

For values far beyond the buckling load the one-term approximation is no longer sufficient. For the calculation of the
deformations in the advanced postbuckling region a series
may be used, the terms of which individually satisfy the
boundary conditions. The coefficients of the series can be
calculated from the condition that the potential energy has
to be a minimum. For an exact calculation many terms had to
be taken into account. But this is not worthwhile, since
the input to the problem will not be given precisely; the
initial imperfections of the plate either are not introduced at all or are treated in a much idealized way.

Fairly useful results are obtained, if for the normal deflection one uses the three-terms expansion

$$w = a_1 \cdot \cos \frac{\pi x}{b} \cdot \cos \frac{\pi y}{b} - a_3 \cdot \cos \frac{3\pi x}{b} \left(\cos \frac{\pi y}{b} - \eta \cdot \cos \frac{3\pi y}{b} \right) \cdot$$

In this it is assumed that the buckles are square. This
corresponds to experience obtained in tests.

Marguerre [3] calculated the two free parameters a_1 and a_3
from the condition that the strain energy has to be a minimum and determined the third free parameter η by trial and
error. In the paper two different criteria are mentioned for
this trial and error procedure, first "that the load carrying capacity becomes as low as possible" and second "that
one obtains the most reasonable course of the load-shortening curve". This method leads to a load shortening curve
which corresponds well to the test results.

Fig.2.13 shows experimental results compared to the theoretical results which have been obtained with the one-term
approximation and the three-term approximation, respectively.
The boundary conditions in both calculation and test are the
same. In both cases the edges of the plate are simply supported and the lateral edge beams can translate freely. It
is unsatisfactory in this comparison that test results are
available only for a limited range beyond the buckling shortening.

Fig.2.14 presents the effective width b_{es} calculated by
means of the three term formula, compared with the test
values obtained by Wagner [1]. The boundary conditions for
analysis and test do not agree. Simple support and freely
sliding edge beams are assumed for the analysis, whereas the
tests were performed with clamped support and non-sliding
lateral edge beams. But when discussing the paper of Wagner
[1] and his staff, it was already mentioned that the authors
considered this transfer of their test results to plates
with other boundary conditions as admissible.

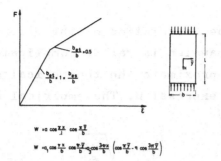

$$W = a \cos\frac{\pi x}{l} \cos\frac{\pi \bar{y}}{b}$$

$$W = a_1 \cos\frac{\pi x}{b} \cos\frac{\pi \bar{y}}{b} - a_3 \cos\frac{3\pi x}{b} \left(\cos\frac{\pi \bar{y}}{b} - \eta \cos\frac{3\pi \bar{y}}{b} \right)$$

Fig.2.12 Schematic load-shortening curve of perfect plates.

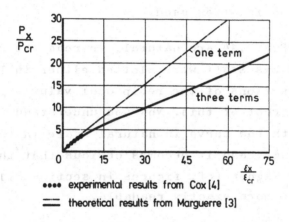

•••• experimental results from Cox [4]

═══ theoretical results from Marguerre [3]

Fig.2.13 Load-shortening curves of nearly perfect plates.

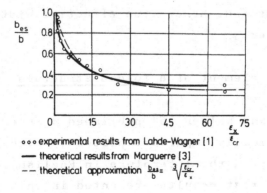

○○○ experimental results from Lahde-Wagner [1]

── theoretical results from Marguerre [3]

── theoretical approximation $\frac{b_{es}}{b} = \sqrt[3]{\frac{\varepsilon_{cr}}{\varepsilon_x}}$

**Fig.2.14 Effective width for stress calculation with nearly
          perfect plates.**

The formula for the calculation of the effective width with the three term expansion is rather complicated, therefore, it was tried to approximate the theoretical curve using a simple analytical expression. The empirical approximation formula

$$\frac{b_{es}}{b} = \sqrt[3]{\frac{\varepsilon_{cr}}{\varepsilon_x}}$$

has proved best. The curve calculated with this formula is also indicated in Fig.2.14. It agrees very well with the test results and is often used.

The scales in Fig.2.14 are natural, whereas the results by Wagner [1] and his staff were plotted either in logarithmic scale or as functions of the reciprocal value $\varepsilon_{cr}/\varepsilon_x$ and /or the square root of this. Now you understand that nothing can be done with the curve in natural scale in the region of small shortenings, and it becomes obvious that the presentation used by Wagner (cf. figures in section 2.1.2) has the advantage to be more exactly readable.

The experimental and theoretical investigations discussed until now, refer to panels the edge beams of which are rigid in bending. In the case of folded structures there are no rigid edge beams and, hence, the effective width of the panels is smaller.

2.1.4   Panel as element of a folded structure

In 1972 Dawson and Walker [5] published a paper "Postbuckling of geometrically imperfect plates", which gives an excellent survey on the literature in English language in this field. The test results presented in this paper were obtained for the most part on cold formed sections; the panels exhibited buckles already in the unloaded state and

their edges did not remain quite straight with increasing
load.

The boundary conditions for the panels making up a section
require compatibility of edge displacements and rotation
between adjacent panels as well as equilibrium of moments
and shear forces. Application of these boundary conditions
is difficult and beyond the means of usual drawing office
design. For simplicity current design specifications assume
simply supported edges for all the elements of a section.
This means that adjacent panels support one another so that
their mutual edges are maintained straight; but that there
are no moments or shear stresses transmitted across the
boundaries.

In the British Specifications [6], the ultimate stress of
the elements of cold formed steel sections is given by

$$\frac{\sigma_{ult}}{\sigma_y} = 0.66 \sqrt[3]{\frac{\sigma_{cr}}{\sigma_y}}$$

determinded by A. Chilver [7] as the lower scatter bound of
plain and lipped, single and double channel sections of
various yield stress. The authors [5] confirm that the design formula of the British Specifications has proved successful in predicting ultimate loads.

Various theoretical methods for calculating the postbuckling
behavior of perfect and imperfect plates are described in
the paper [5] and the results are compared. In practice no
plate is perfectly flat, and buckles start forming from the
onset of loading. It might seem then, that σ_{cr} is of only
academic interest. However, this is not appropriate; σ_{cr}
provides a useful parameter in the non-dimensional analysis
of plates, and the corresponding eigenmode indicates the
most dangerous imperfection shape.

For the latter reason, and for sake of convenience, in the
analysis the initial out-of-flatness is assumed to have the
shape of this eigenmode. For the amplitudes of the initial
imperfections three different formulas have been proposed:

$$\frac{a_0}{t} = \alpha \, ,$$

$$\frac{a_0}{t} = \beta \cdot \sqrt{\frac{\sigma_y}{\sigma_{cr}}} \, ,$$

$$\frac{a_0}{t} = \gamma \cdot \frac{\sigma_y}{\sigma_{cr}} \, .$$

α, β and γ are constants, which are chosen so that the
experimental and theoretical ultimate loads agree as well as
possible. All the curves $\alpha = 0.2$, $\beta = 0.2$ and $\gamma = 0.2$
fit the test results reasonably. The best fit is given by
the γ-curve. Because of this

$$\frac{a_0}{t} = 0.2 \frac{\sigma_y}{\sigma_{cr}}$$

is recommended as the generalized imperfection paramter to
be used as the basis for drawing up design curves for cold
formed sections.

The ultimate load is defined as the load at which the membrane stress reaches the yield limit σ_y. Fig.2.15 shows
the good agreement between calculated ultimate loads and
measured collapse loads. For each panel b/t or σ_y / σ_{cr}
there is only one point in Fig.2.15 since each panel has
only one collapse load.

The effective width not only depends on the plate dimensions
and material, but also on the applied load. Therefore, they

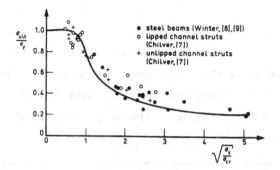

Fig. 2.15   Experimental and theoretical collapse loads of imperfect plates .

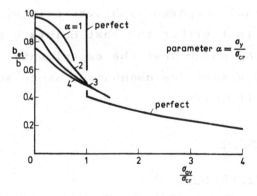

Fig. 2.16   Effective width for stress calculation with imperfect plates .

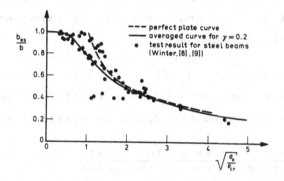

Fig. 2.17   Effective width for buckling calculation with imperfect plates .

must be represented by two curves for each plate, one for b_{es} and one for b_{et}. For the effective width b_{es}, needed in stress calculation, Dawson et.al.[5] propose to replace the family of curves, calculated for a number of plates, by the average curve, shown in Fig.2.16. As abscissa the ratio

$$\sqrt{\frac{\varepsilon_e}{\varepsilon_{cr}}} = \sqrt{\frac{\sigma_e}{\sigma_{cr}}} = \sqrt{\frac{\text{edge stress, averaged over the family of curves}}{\text{buckling stress of the panel}}}$$

is used. The curve is compared with test results. Only moderate agreement is obtained as could be expected due to the rough averaging process.

For the effective width b_{et}/b which is needed, in the buckling analysis for sections, the family of curves is retained. It is shown in Fig.2.17. As abscissa the ratio

$$\frac{\sigma_{av}}{\sigma_{cr}} = \frac{\text{average stress of the panel}}{\text{buckling stress of the panel}}$$

was chosen. The curve parameter is

$$\alpha = \frac{\sigma_y}{\sigma_{cr}} = f\left(\frac{b}{t}\right).$$

This diagram was not compared with test points. From this it can be concluded that either the test points scattered too much for a publication or that the experimental confirmation of the calculated values was abandoned, since such tests would be rather difficult.

2.2 Welded box beams

2.2.1 Preliminary remarks

In the following, experimental and theoretical investigations on steel columns the cross sections of which are thin-walled

rectangular boxes, are discussed. Research work was conducted by K. Klöppel and collaborators [10, 11] at the
Technische Hochschule, Darmstadt. An excellently equipped
experimental laboratory was at disposal for the tests and
a computer for the calculation. Research work was aimed at
experimentally determining the ultimate load and placing a
method of calculation at the designing engineer's disposal,
that allows him a predetermining of ultimate loads, with
reasonable amount of computation. The object was achieved.

2.2.2 Experimental investigations

Experiments were performed not with models but with full-
size types, because only then it was guaranteed that the
initial imperfections and initial stresses occurring could
be incorporated in actual magnitude and shape. Fig.2.18 shows
a schematic sketch of the specimens. They consist of St 32-2
steel having a yield limit of 240 N/mm^2. The corners are
welded.

Special care was placed on uniform load introduction. At the
ends of the box on both sides of the panels strong cover
plates were screwed on. The bolt holes were so large that
the cover plates could be installed in the test device, so
that they fitted along their whole length against the compression plates of the device. Only then, high-tensile screws
were tightened.

The initial imperfections as well as the initial stresses
were measured and then included in the calculation in the
approximate full magnitude, but with idealized shape. This
effort proved to be necessary in order to obtain agreement
between the experimental and theoretical results.

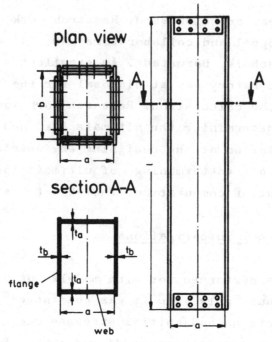

plan view

section A-A

flange

web

Fig.2.18 Test specimen .

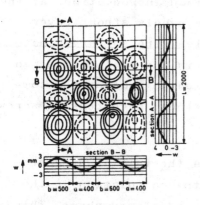

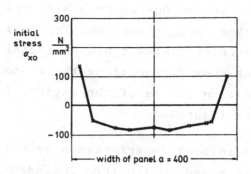

Fig.2.19 Measured initial
imperfection pattern .

Fig.2.20 Measured initial
stresses .

Fig.2.19 shows measured imperfections, which are approximately affine, to the buckling pattern of the perfect box. The largest amplitude equals half the wall thickness. The imperfections were distributed so regularly in a few cases only.

The initial stresses of the specimens could not be measured by a non-destructive method. They were determined for a reference beam with square cross section. In order to make the initial stresses apparent the beam was sawed into small, nearly stress-free pieces. Fig.2.20 shows measured initial stresses. It can be recognized that high tensile stresses act in the corners of the box and that compressive stresses, which are almost evenly distributed over a wide range act in the center region of the panel.

The load was applied stepwise, using a power controlled hydraulic $500 \cdot 10^4$ N universal compression-tension loading device. The approach of the ultimate load was indicated by the phenomenon that the deformations did not stop when a load increment was finished, but rather crept slowly to a final value. A further characteristic for approaching the limit of load carrying capacity was the crackling tearing-up of tinder in the regions of highest stresses.

These crack lines can be recognized at the edges of the buckles in Fig.2.21. The column K 23, shown here, is characterized by thin webs with large imperfections, and sturdy flanges, with negligibly small imperfections. The load acts centrally. Already with relatively low load, the buckles in the webs have grown large and affected the flanges. Fig.2.21 shows that the webs and the flanges are buckled in the same rhythm, but that the buckles in the flanges have smaller amplitudes.

32

Fig.2.21 Postbuckling pattern of the centrally loaded box beam K 23 .

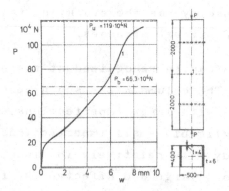

Fig.2.22 Load deformation curve of the centrally loaded box beam K 23 .

Fig.2.22 shows the load deformation curve of the buckling
process. The local buckling load calculated for the perfect
column amounts to $P_b = 66.3 \ 10^4$ N; it is plotted somewhat
above the middle of the diagram. The abscissa shows the increment of the deflection in the middle of a web buckle.
The amplitude of the web buckle is finitely large from the
beginning: the buckle depth increases non-linearly already
with small loads. With one third of the buckling load of
the perfect column, obviously the flanges begin to buckle
too. This reduction of the buckling load can be explained
by two facts: Firstly, the load was mainly carried by the
flanges, because the webs withdraw from load increase by
deepening the buckles. and secondly, high initial compressive stresses due to welding the edges, existed in the
middle of the flanges.

After the flanges have been included in the buckling process, the slope of the load deformation curve again becomes
steeper, because the deflections in the panels become so
large that they bring about the effect of supporting by the
membrane forces. (Please, remember the schematic sketch,
Fig.1.2 showing a plate in the postbuckling region, which
is presented in the introduction.) The magnitude of the
restraining membrane forces depends on, whether the plate
has rigid edge beams or is flexibly supported in its plane
as in the case of box beams. But nevertheless this supporting effect always exists in plates.

Passing the theoretical buckling load causes no kink on the
load deflection curve. It is without significance for the
buckling process, since the longitudinal stress in the box
is not evenly distributed; the stress is higher in the corners of the box and is lower in the middle of the buckled
panels.

When the loading approaches the ultimate load, the slope of
the curve again becomes flatter, because the elastic limit
is exceeded locally.

Fig.2.23 shows postbuckling patterns of the square column
K 25 with a wall-thickness that is uniform all around. This
column had negligibly small initial buckles in all panels.
The box beam is centrally loaded. On the left hand picture,
you can notice square buckles which extend in exact regularity over the whole circumference and the whole length.
On the right hand, you can see the column after exceeding
of the ultimate load. After folding over, of the edges, for
which a shortening of more than 2 cm was required the welding seams tore off. Consequently, the load dropped.

Fig.2.24 shows the load deflection curve of the same box
beam. As the abscissa the normal deflection in the middle
of a buckle is plotted. Up to somewhat more than half the
local buckling load of the perfect structure the depth of
the buckle remains negligibly small. Then the curve bends
over in a short transition piece and runs on with flatter
slope. It can be learned from this curve, that the buckles
have jumped in simultaneously in all panels at half the
theoretical buckling load. The reduction of the buckling
load is due to the fact that initial compressive welding
stresses existed in the center region of the panels.

The further course of the load deflection curve is principally the same as for the box beam, Fig.2.22. The slope of
the curve increases, when the buckles grow deep, because the
restoring membrane forces become effective. The theoretical
buckling load is not a clearly marked point in the buckling
process, because the corresponding stress is scarcely present in the column. The stresses in the corners lie above

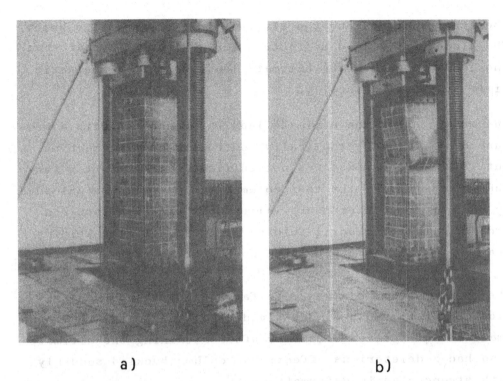

a)   b)

Fig.2.23   **Postbuckling patterns of the centrally loaded box beam K 25 .**

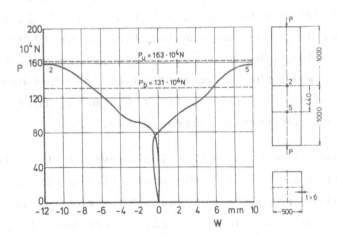

Fig.2.24   **Load shortening curves of the centrally loaded box beam K 25 .**

the average stress and the stresses in the panels lie below.
The slope of the load deflection curve becomes **flatter when**
the load approaches the ultimate load, because the elastic
limit is exceeded locally.

The attaining of the ultimate load was connected with a pressure drop in the hydraulically controlled testing machine,
but this load decrease was not sufficient to prevent a further shortening of the test columns unless, it was unloaded
by opening the oil pressure valve. Only after further controlled shortening until folding of the edges, the welding
seams tore off, as can be seen on Fig.2.23, right hand frame.

The two box beams discussed so far had low slenderness ratios. Fig.2.25 shows a more slender column. In this case the
small eccentricities occurring after attaining the ultimate
load had a deleterious effect: The column buckled suddenly
with strong plastic deformations.

The last box beam treated in this chapter is a very slender
one, which was compressed eccentrically. Fig.2.26 shows the
column K 35 after attaining the ultimate load. It can be
noticed that local and column buckling occurred simultaneously. The local buckles can be recognized well by means
of the grid. When the ultimate load was reached one sharp
kink appeared in the lower half of the beam. This can be
noticed below the strain gauges in Fig.2.26.

Fig.2.27 shows the load deflection curve of the column K 35.
The deflection and, hence, the eccentricity of the load increases nonlinearly with the load. This results in an earlier buckling of the more strongly compressed flange. It is
noticeable that for this slender eccentrically loaded box
beam, the ultimate load is the column buckling load. This
can be seen from the fact, that the load deflection curve

Fig.2.25 Postbuckling patterns of the slender centrically loaded box beam K 34 .

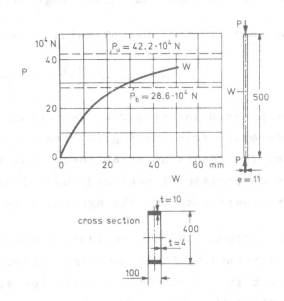

Fig.2.26 Postbuckling pattern of the very slender excentrically loaded box beam K 35 .

Fig.2.27 Load deformation curve of the very slender excentrically loaded box beam K 35 .

asymptotically approaches the ultimate load. Please remember
and compare Fig.1.5 in the introduction.

2.2.3 Theoretical investigations

As already mentioned the initial stresses were measured not
on the specimens themselves but on a square box beam that
had to be destroyed for the measurements. In the calculations the initial stresses were given by an approximate formula, based on measurements and on theoretical considerations.
In the welding seams of the edges high tensile stress peaks
occur which are equilibrated by an extended compressive stress
field in the middle region of the panels, Fig.2.28. The calculation takes only residual m e m b r a n e stresses into
account. The stress distribution, shown in Fig.2.28, decays to
zero at the upper and lower end of the box beam. But the columns are so long that these edge effects may be neglected.

The initial stresses following from the formula

$$\sigma = A \cdot \cosh \frac{2 \pi y}{l} + B \, \frac{2 \pi y}{l} \cdot \sinh \frac{2 \pi y}{l}$$

are plotted in Fig.2.28a. The parameters A and B are
determined from the edge stress σ_e and from the condition
that the resulting axial force must be zero. Fig.2.28b shows
the comparison of measured and calculated initial stresses
for square buckles; the agreement is good.

The influence of the residual stresses is of great interest
in determining local buckling deflections and of little interest in determining ultimate loads, since due to local
buckling the compressive stresses in the center of the panels are largely reduced, when the ultimate load is app-
roached.

The shape of the initial imperfections is assumed to be affine to the buckling pattern of the perfect box beam.

Fig.2.29 shows once again the measured load deflection curve for the square column K 25 and, additionally, four calculated curves. The prebuckling amplitude was measured to be 0.3 mm : this amplitude and an imperfection pattern that is affine to the buckling mode of the perfect column was introduced in the calculation. If, moreover, residual stresses with a stress peak of σ_e = 200 N/mm^2 in the box corners are taken into account, the lowest of the theoretical load deformation curves is obtained. This theoretical curve agrees well with the experimental curve up to the vicinity of the ultimate load P_u = 163·10^4 N : in the experiment obviously plasticity phenomena occur, which have not been considered in the calculation.

In the next higher theoretical curve, the initial stresses, but no imperfections, are taken into account; this curve also corresponds to the experimental curve to some extent. In the next higher curve the imperfections are considered, but no residual stresses, and in the top curve neither imperfections nor initial stresses are taken into account. The two upper curves do no longer agree with the test results.

The agreement, between the theoretical load deflection curve, calculated with consideration of the residual stresses, and the load deflection curve, based on experimental results, confirms fortunately, that the relations were adequately considered, in the theory. However, it remains undecided, whether the successful investigations on this example mean, that any box beam can be satisfactorily analysed by the method applied here. It must be considered that this column had a low slenderness ratio, that the

40

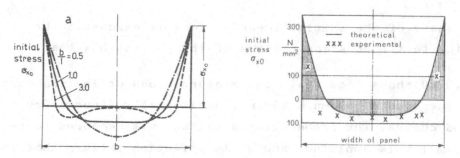

Fig.2.28 Theoretical and experimental initial stress .

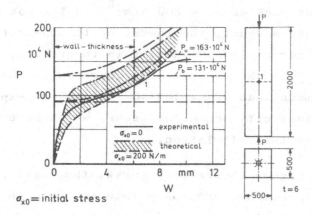

σ_{x0} = initial stress

Fig.2.29 Theoretical and experimental load shortening
curves of the centrally loaded box beam K 25 .

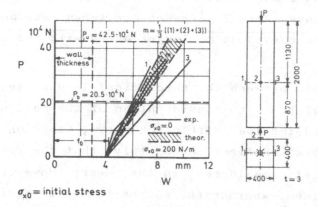

σ_{x0} = initial stress

Fig.2.30 Theoretical and experimental load shortening
curves of the centrally loaded box beam K 27 .

imperfections are small in this example and that the initial stress was chosen so that the experimental and theoretical results agreed as well as possible.

Fig.2.30 shows calculated and measured load deflection curves of a square column with low wall thickness and large imperfections. The calculations were performed with a welding initial stress of 200 N/mm^2. Agreement between experimental and theoretical values is extremely good, especially when comparing the calculated deformations with mean measured ones.

Fig.2.31 shows the measured and the calculated load deflection curve of the slender eccentrically loaded column K 35. The load deflection curve is calculated with neglection of the initial imperfections and the initial stresses. Hence, the calculated deflections are essentially smaller than the measured ones. The local buckling load $P_b = 28.6 \ 10^4$ N is determined from considering the elastic lever arm; but since the calculated eccentricity is smaller than the actual one, the calculated buckling load will be overestimated. Nevertheless the ultimate load is considerably higher than this overestimated theoretical local buckling load.

For calculating the ultimate load, one must at first decide how the ultimate load is to be defined mathematically. In Fig.2.32 different limits of ultimate load are compared to each other for a box beam with imperfections.

I.   The average stress at which the depth of buckles attains the limit f = 0.0100 a .

II.   Buckling stress, with reduction in the plastic region.

III. The average stress at which the depth of buckles attains the limit f = 0.0125 a .

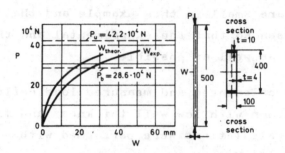

**Fig.2.31 Theoretical and experimental buckling deforma-
tions of the box beam K 35 .**

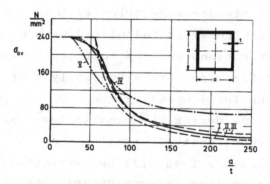

**Fig.2.32 Theoretically defined ultimate stress as function
of the width-to-wall-thickness ratio .**

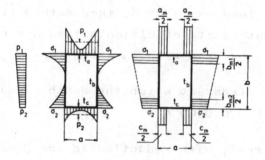

Fig.2.33 Effective width for bending calculation .

IV.   The average stress at which the membrane stress peak
      attains the yield limit.

V.   The average stress at which the total stress (membrane
      plus bending stress) attains the yield limit.

This fine subdivision can be condensed in two groups: the
lower group I, II, III contains the loads at which buckling
deformations occur; the upper group IV, V contains the loads
with which the elastic limit is attained. It is noteworthy
that, for thin-walled columns the load at which the yield
limit is attained, is five times as high as the buckling
load. It would be uneconomical, if for thin-walled columns,
the load reserves, existing in the postbuckling region,
would not be utilized.

In the following the carrying capacity limit IV is considered to be decisive, that is, those loads are calculated
at which the membrane stresses attain the yield limit. Postbuckling calculations require the application of a nonlinear
plate theory and are therefore so expensive that they, in
general, cannot be performed by the design engineer. Consequently the problem arises how considerable parts of the
calculations can be anticipated for the practical case, and
how the influence of local buckling on the column buckling
of a centrally or eccentrically compressed box beam can be
taken into account.

In discussing the postbuckling behavior of plates we have
understood that the center region of the plate with increasing load withdraws from taking high compressive stresses.
This led to the concept of an effective width.

In a corresponding manner, for box beams an effective cross-
section is introduced. Fig.2.33 at the left hand side illus-

trates the stress distribution in a box beam under combined
longitudinal compression and bending. Instead of the linearly distributed stresses p_1 and p_2 a nonlinear stress
distribution occurs, with the peak stresses σ_1 and σ_2 at
the corners and almost vanishing stresses in the middle of
the panels. At the right hand side you see the effective
equivalent cross section, which is se defined that under
the same loading, by applying the linear bending theory,
the same stresses σ_1 and σ_2 result at the edges as for
the full cross section when applying the nonlinear plate
theory.

In calculating the effective cross section, the imperfections are taken into account according to empirical values
as in calculating the buckling loads of the box beams. The
initial stresses are neglected, since the compressive
stresses in the middle of the buckles will be largely reduced when attaining the ultimate load.

Since the effective cross section of a box beam is principally the same as the effective width of a panel, one has
to distinguish also between the effective cross section for
the calculation of stress, and the effective cross section
for the calculation of buckling loads. The effective cross
sections can be determined by computing the effective widths
of the individual panels forming the box beam. At Klöppel's
institute the effective widths of the panels were computed
in advance and presented in diagrams. Fig.2.34 shows as an
example the effective width for s t r e s s calculation
for the panels of a square box beam with constant wall-
thickness under central loading. As abscissa the average
compressive stress was chosen. Curve parameter is the ratio

$$\frac{a}{t} = \frac{\text{panel width}}{\text{wall thickness}} .$$

The more thin-walled the box, the smaller the effective
width. The ultimate load is attained, when the stress

$$\sigma = \sigma_{av} \cdot \frac{a}{a_m}$$

has reached the elastic limit $\sigma_e = 240$ N/mm^2 .

For comparison, but without practical importance, the curve
for the buckling stress of the perfect panel is also plotted
in the diagram. One recognizes that for thin-walled panels
the buckling load is smaller than the ultimate load. When
a/t = 70 , buckling and ultimate load are equal. For thick-
walled panels the buckling load was calculated taking into
account a reduction due to plasticity. In an intermediate
region it is larger than the ultimate load. For very thick-
walled panels, if the whole panel width carries uniformly,
buckling load and ultimate load are practically equal.

The utilization of the load carrying capacity in the postbuckling region means an enormous increase of profitability.
Let us consider as an example a square box beam with the
width-to-wall-thickness ratio a/t = 200 . The panels buckle,
when $p_{av} = 14,50$ N/mm^2 and attain the elastic limit, when
$p_{av} = 57,00$ N/mm^2 : hence, the column can be loaded four
times as high, when the elastic limit of the membrane
stresses in the buckled panel is considered as being decisive rather than the buckling load.

But such an increase of the ultimate load does not neces-
sarily mean that the admissible load may increase by the
same factor. For the design of a structure the engineer not
only should know the ultimate load, but should also know
what margin of safety is adequate in the case at hand. Let
us use the opportunity to lodge some general remarks on the
margin of safety. It depends on:

- First, the danger class. If human life will be endan-
  gered by a failure of the structure, safety must be
  higher than for a simple toy.

- Second, the behavior of the structure after having
  reached the ultimate load. For an axially compressed
  thin-walled circular cylinder, the load of which drops
  after buckling, one will provide a higher safety margin
  than for an elastic column the equilibrium state of
  which is stable in the initial postbuckling region.

- Third, the determination of the ultimate load. In former
  days the computation techniques were less developed than
  in our days. For instance the load carrying reserves of
  the postbuckling region were not taken into account
  explicitely. So the safety margin for buckling plates
  could be extremely small, for instance $S = 1.25$ in
  civil engineering. But when the ultimate load is defined and calculated as the load at which the membrane
  stresses reach the yield stress, larger safety margins
  must be used.

After this little digression we return to our example, i.e.
the thin-walled box beam, and we remind you that the ultimate
load defined according to Fig.2.34 is four times as high as
the buckling load of the panels. However, since in this ultimate load the reserve capacity of the postbuckling region is
included, the margin of safety must be larger than for a design based on the buckling load of the panels. This might
result in an increase of the admissible load by a factor of,
say three, only.

For square columns with constant wall-thickness and central
loading, one diagram for the effective width is sufficient.

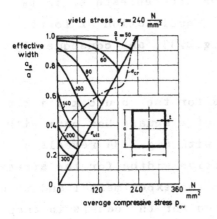

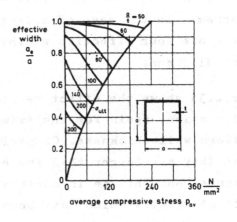

Fig.2.34 Effective width for stress calculation .

Fig.2.35 Effective width for buckling calculation .

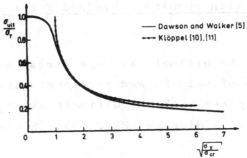

Fig.2.36 Comparison of collapse loads $\sigma_y = 250$ N/mm^2.

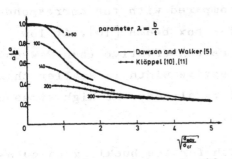

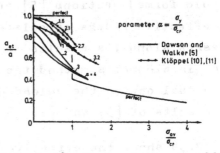

Fig.2.37 Comparison of effective width for stress calculation .

Fig.2.38 Comparison of effective width for buckling calculation .

If the box is not square and / or the wall-thicknesses are
different, then the effective widths are represented in two
diagrams. In the case of eccentrical application of force,
there are four effective widths, Fig.2.33, and consequently
four diagrams.

Fig.2.35 shows the effective widths for the b u c k l i n g
calculation, again for the example of a square box beam with
uniform wall-thickness. Comparison with Fig.2.34 reveals
that they are larger than the effective widths for the stress
calculation. This is in contrast to the experimental and theoretical results which have been given for the panels in chapter 2.1 .

2.2.4. Comparison with results, obtained for cold formed sections

Fig.2.36 presents the ultimate average stress of panels,
which are elements of cold formed sections, given by Dawson
et al. in [5], compared with the ultimate average stress
deduced from Klöppel's diagram, Fig.2.34. They are in excellent agreement.

In Fig.2.37 the effective widths for the stress calculation
of cold formed sections [5] are compared with the corresponding effective widths calculated for box beams [10]. At low
stresses Klöppel's results are more conservative than Dawson's
since in Klöppel's method the effective width is smaller than
the actual one in the unloaded state already. At high stresses
the results of [5] and [10] merge.

Fig. 2.38 shows the effective width for the buckling calculation. The agreement is unsatisfying. That was to be expected,
since Klöppel in contrast to the other scientists indicates
a higher value for the tangent than for the secant stiffness.

2.3 Open profiles of cold formed steel

2.3.1 Preliminary remarks

In the following, experimental and theoretical investigations
are discussed, which were published by K. Klöppel [12] and
his staff at the Technische Hochschule Darmstadt in 1971.
The test results are valuable and can be used in design or
can serve as a basis for theoretical studies. The theoretical
investigations in this paper do not meet the difficulty and
complexity of the problem.

2.3.2 Experimental investigations

Test specimens with the cross section, shown in Fig.2.39,
and different lengths were used. The support on both ends
with good approximation corresponds to the following boundary conditions, Fig.2.40a .

$$
\begin{array}{llll}
\text{Bending moment} & M_\xi & = & 0 \\
 & M_\eta & = & 0 \\
\text{Angle of rotation} & \vartheta & = & 0 \\
 & \vartheta' & = & 0 \\
\text{Displacement} & u & = & 0 \\
 & v & = & 0 \; .
\end{array}
$$

In order to demonstrate the influence of warping of the end-
sections on the load carrying behavior in some cases the
support of the test specimens was modified so that the
warping moment became zero instead of $\vartheta' = 0$. This could
be easily achieved by chamfering the flanges of the test
specimens, as shown in Fig.2.40b and not screwing them to
the base plate. The remaining cross section, a simple angle,
is warpfree.

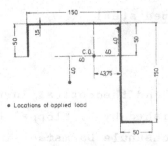

Fig.2.39 Cross section of the column .

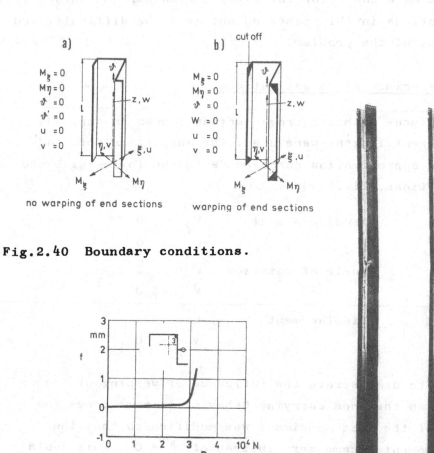

Fig.2.40 Boundary conditions.

postbuckling patterns

**Fig.2.41 Buckling and postbuckling behavior of an eccen-
trically loaded long column L = 4000 mm.**

The ultimate load was defined as being the load at which
either a further load increment could not be carried by the
test specimen or at which one of the unstiffened flanges of
the column, i.e. one of the panels with a free longitudinal
edge, failed by local buckling. It should be mentioned that
these two criteria are not equivalent; in some cases the
unstiffened flange failed owing to a local buckle, the
column however withstood a further load increase.

Only a few examples are chosen from the abundance of data
offered in the paper. We choose the longest column which has
a length of 4000 mm, and the shortest one, which is 1000 mm
long.

L o n g   c o l u m n s   fail by bending and twisting
simultaneously. But this is generally preceded by a deformation of the cross section due to local buckling.

Fig.2.41 left hand side, shows the load deflection curve for
load application at a point near the corner. In the schematic
sketch at the top the load axis is marked by a thick point.
The position where the deflection was measured and the positive direction of measurement are marked by an arrow with a
circle in the middle, i.e. the symbol for a displacement gage.

The buckling load that amounts nearly to $3 \cdot 10^4$ N can be
recognized by a distinct kink in the load deformation curve.
Prior to attaining the buckling load, deflections were practically nonexistent; that points to the fact that the test
columns were very carefully manufactured.

After buckling the slope of the load deformation curve is
very steep. Hence, for this long column there are practically no load carrying reserves in the postbuckling region.

Fig.2.41, right hand side, presents the unloaded column
after the buckling test. In the postbuckling region the
yield limit was attained over a large range of the section.
Hence, the postbuckling deformations were preserved. From
this figure it can be learned that the strut was long enough
to buckle as a column.

For loading at the two other load application points indicated in Fig.2.39, no load deformation curves were recorded
for the long strut, probably because the deformations were
so large that the measuring device would have been overshot.

At the s h o r t   c o l u m n   failure was caused by
local buckling; buckling or torsion of the whole column was
of no importance.

Fig.2.42, left hand side, shows two load deformation curves
of the centrally loaded strut. The buckling load amounts to
nearly $4 \cdot 10^4$ N .

In the postbuckling region the deflections increase strongly,
but less than at the long column; one deflection increases
in positive, the other in negative direction. That leads to
the thought that the section would twist, as the long column
did. But this conclusion would be wrong. On the plastically
buckled column, shown on the right hand side, one can recognize that the measured deflections were buckling amplitudes
of the panels one in inward and the other in outward direction.

Attaining the ultimate load is marked by buckling of that
unstiffened flange, which is directed inward and, hence,
is more strongly loaded than the other one. The buckle on
this flange grew so large that no further load increase was

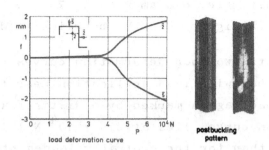

Fig.2.42 **Buckling and postbuckling behavior of a centrally loaded short column L = 1000 mm.**

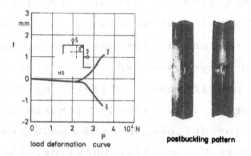

Fig.2.43 **Buckling and postbuckling behavior of an eccentrically loaded short column L = 1000 mm.**

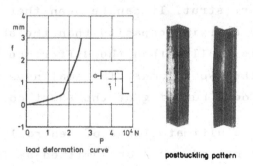

Fig.2.44 **Buckling and postbuckling behavior of an eccentrically loaded short column L = 1000 mm.**

possible. The ultimate load amounts to $7 \cdot 10^4$ N and consequently is 75 % higher than the buckling load.

Fig.2.43 presents two load deformation curves of an equal test specimen, but with load application near the corner. The buckling load, again marked by a distinct kink in the load deformation curve, lies at $2.5 \cdot 10^4$ N ; it is essentially smaller than for the centrally loaded strut, shown in Fig.2.42. The ultimate load is also smaller; it lies at $3.5 \cdot 10^4$ N . The final failure is triggered by a buckling of the loaded corner, see photo on right hand.

Fig.2.44, left hand side, illustrates the load deformation curve of a third equal test specimen, this time with eccentric application of load at the most space centered point. The buckling load is yet smaller; it lies at $1.6 \cdot 10^4$ N . In this case the unstiffened cantilevering flanges are loaded most strongly. They already withdraw at a rather low load. Hence, the deflection at the measurement point has already increased to 0.5 mm till buckling. The ultimate load lies at $2.5 \cdot 10^4$ N and is characterized by a buckling of the unstiffened inner flange.

Fig.2.45 presents a comparison of the ultimate loads of the long and the short strut. It can be seen that the long strut has a lower load carrying capacity than the short one. But the difference is smaller than the difference between the Euler loads of the two columns. The smallness of the difference is due to local buckling of the short column.

Fig.2.46 shows the ultimate loads of centrally loaded struts with warping restrained and / or admitted as function of the column length. It can be seen that the ultimate load of long columns is decreased up to half the value by allowing the end section to warp. For short columns the influence of the

boundary conditions is smaller, since short columns do not
tend to twist, but rather fail by local buckling. These
test results, however, are only indications and no reliable
design data. Their number is much too low to admit generally
valid conclusions.

2.3.3 Theoretical investigations

Theoretical investigations on the postbuckling behavior of
open profiles which can fail in a torsional mode are much
more difficult and laborious than the postbuckling calculations for the rectangular box girder. As far as we know,
such calculations have not been performed yet. But such
profiles are used in the engineering practice, and therefore the designer at least must have approximate formulae
with which he can estimate the load carrying capacity to
some extent.

Klöppel and his staff computed the ultimate loads of their
test specimens with the available calculation methods and
compared the results mutually and with the experimental
values. Figs.2.47 and 2.48 show these comparisons for the
short and the long strut, respectively. Before discussing
these diagrams the meaning of the different curves is explained. They will be subdivided in three groups.

The first group contains column buckling. These curves are
plotted with continuous lines:

(a) means the ultimate load, resulting from the ω -method [13],
    legally prescribed in Germany for the design of columns,

$$P = \frac{S_0 \cdot \sigma_{ult}}{\omega} \quad ,$$

where S_o is the cross section aera less bolt holes
and ω is a reduction factor which depends on the material

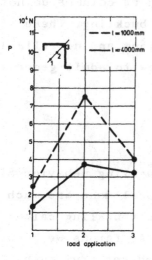

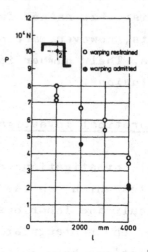

Fig.2.45 Ultimate loads
of columns with different
lengths.

Fig.2.46 Ultimate loads
of columns with and with-
out warping.

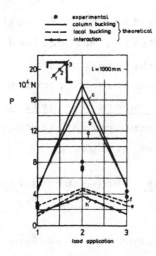

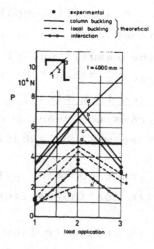

Fig.2.47 Experimental
ultimate loads compared
with different theoretical
loads for the short column
L = 1000 mm.

Fig.2.48 Experimental ultimate loads compared with
different theoretical loads
for the long column L = 4000 mm.

and the slenderness of the column. For the short column $\omega = 1.09$, for the long column $\omega = 2.44$. In this calculation besides the local buckling, also the eccen-
tricity of load application and the influence of torsion
are neglected.

(b) differs from (a) only in that the weakening of the cross
section aera by the bolt holes is neglected, but the eccentricity of the load application is taken into account.

(c) is the ultimate load, which results from a s t r e s s
calculation according to second order theory considering
bending and twisting deformations. Thus it was assumed
that the column section remained preserved, that means
local buckling was neglected, as was weakening of the
cross section by bolt holes.

(d) means the   b u c k l i n g   l o a d , calculated for
torsional-flexural buckling. Local buckling and the
weakening of the cross section by bolt holes are neglected. This load has been evaluated for the long column
only.

The second group contains local buckling. These curves are
plotted with dashed lines:

(e) is the lowest buckling load in one of the panels. In
this buckling calculation, the edges are understood as
hinges remaining straight. Each panel is assumed to
buckle with the pattern, that results in its lowest
buckling load. Initial imperfections are not taken
into account.

(f) means the local buckling loads of the panels, calculated
under consideration of interaction of the different

panels. The corner lines remain straight. Compared to
(e) this method is closer to reality in two respects:
Firstly the mutual elastic restraint of the panels is
taken into account and secondly all panels buckle with
the same wave length.

(g) differs from (f) only in that the corners do not remain
on straight lines. This mode was evaluated only for the
long column.

The third group contains interaction of column and panel
buckling. There is only one curve, marked by small circles:

(h) represents the ultimate loads which result from the
American Standards. Postbuckling of the panels is taken
into account by introduction of the effective width.
The torsional buckling of the whole column is neglected.

In Fig.2.47 you see the theoretical and experimental results
for the short strut. All calculations performed for column
buckling, neglecting the local buckling, yield too high values for the ultimate load. Hence, the diagram demonstrates
the failure of the strut is induced by local buckling. The
curves for panel buckling and for the interaction of panel
and column buckling, lie near together. They all yield reliable design data, but are in most cases too conservative.
The discrepancy is especially large for the centrally loaded
structure, point 2 .

Fig.2.48 presents the results obtained for the long strut.
Here the column buckling loads are much smaller than those
shown in Fig.2.47 for the short strut. But with one exception they are still higher than the experimental ultimate
loads. The calculated panel buckling loads, in some cases,

turn out to be also higher than the measured values; hence, they are no longer reliable design data, as they were for the short strut.

The only theoretical loads which prove to be equal to or smaller than the measured values are those calculated according to the American Standards. This statement is valid for the short and for the long strut. But in spite of this the American Standards also may not be considered as a satisfying solution of the problem. This assertion can be proved by the following considerations: Regarding only the loads of the centrally loaded struts, point 2, one finds that for the short strut the calculated load is half the measured one and that for the long strut the calculated and measured values agree; from this it may be extrapolated, that for still longer struts the calculated loads would become higher than measured ones and consequently worthless for design purposes. This assumption is confirmed by the fact that long struts fail by torsion and that this buckling mode is not within the scope of the American Standards.

Concluding, it can be stated that until now no satisfactory calculation method exists for the determination of the ultimate loads of long open cold-formed sections, when postbuckling of the panels and column buckling of the whole strut interact. This statement is indeed true, but it may be remarked, that large effort would be required for a suitable theoretical solution. This effort however, is perhaps not justified since long columns with open sections are not suited for transfer of high compressive forces. A good designer will not use them for this purpose.

For the long centrally loaded column, treated in [12] the average stress at ultimate load is only 20 % of the yield stress. This vslue is too low for an economic construction.

2.4   Arches under bending

A circular arch, Fig.2.49a, whose bending stiffness in the
plane of its curvature is essentially larger than in a plane
perpendicular to it, and whose torsional stiffness is not
too great, can buckle under either tensile or compressive
loading. You may be able to satisfy yourselves on that phenomenon by testing a model arch cut out from cardboard or
from a piece of foil. This threedimensional buckling is so
complicated that there is at this time only a buckling calculation [14, 15] without a postbuckling calculation. However, we can report on a buckling test where the investigation has proceeded to the postbuckling region.

Fig.2.49b shows the load deformation curve of a circular
arch, made of flat steel, under tension. One sees, that in
the prebuckling region the deformation increases linearly
with the load. When the buckling load is reached, the equilibrium configuration becomes clearly unstable. The carriage
moves away about 75 mm at constant load. Then the structure
finds a new stable equilibrium configuration. The load could
again be increased till it reached twice the value of the
buckling load. At this stage the test got interrupted as
probably the capacity of the test device was exhausted. The
slope of the load deformation curve is steeper in the postbuckling region than in the prebuckling region.

Both the ends of the specimen were restrained against out-
of-plane deflections. Fig.2.50 presents a photograph of the
test specimen shortly before buckling. From this figure it
can be learned that the deformation pattern was symmetric
and that the out-of-plane deformations in the middle of the
arch have become large, even before buckling.

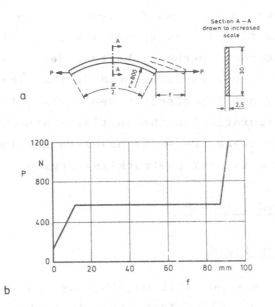

Fig.2.49   Load deformation curve of an arch loaded by tension.

Fig.2.50   Prebuckling deformations before the snap-through.

The measured buckling load was 5 % greater than the calculated one. This slight discrepancy might be attributed
to experimental errors; perhaps the test device was not
entirely friction-free. It is remarkable that the buckling
load is obviously not imperfection sensitive, although the
equilibrium configuration in the initial postbuckling region
is unstable. Thus there are still unsolved problems for the
researchers in the initial postbuckling region.

2.5 Web subjected to shear

2.5.1 The tension field

In the following, a paper will be discussed which was published by H. Wagner [16] in 1929. It illustrates a structural element developed by the author during his occupation
in the Rohrbach Metallflugzeugbau. Probably this was the
first time that the postbuckling behavior of structures was
systematically investigated and methodically utilized.

The web of the beam serves to carry the transverse force and
is therefore mainly loaded by shear. As it is well known,
shear is equivalent to a tension-compression load in orthogonal planes, which are inclined by 45° to the direction of
the shear, Fig.2.51a. When the shear buckling load is exceeded, it implies that the compressive forces have exceeded
their buckling limit and that a further increase in the transverse force P can only be taken up by increasing tension
forces.

In the case of a very thin web and a relatively high transverse force, in the buckled web the compressive forces can
be neglected, compared to the tension forces. Essentially,
the web only acts in tension; it has become a tension field.

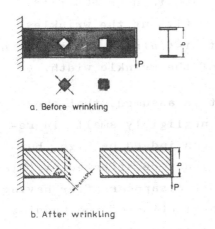

a. Before wrinkling

b. After wrinkling

Fig.2.51 Internal stresses in the tension field.

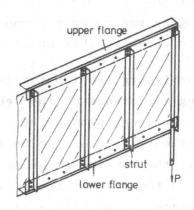

upper flange

strut

lower flange

Fig.2.52 Structure of a tension field beam.

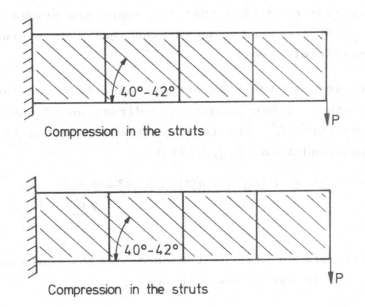

40°–42°

Compression in the struts

40°–42°

Compression in the struts

Fig.2.53 Deformation of the struts and flanges.

After the buckling of the web, the load can be still considerably increased, for instance 100-fold or 500-fold,
without the beam breaking down and without the wrinkles
forming an exceedingly strong out-of-flatness. In this, the
wrinkle depth reaches about 5 % of the wrinkle width.

When designing tension fields, it is assumed, that the
bending resistance of the web is negligibly small. In reality the local bending stresses can indeed be high, but
the ultimate load of the beam will not be affected by this,
because the bending stresses almost disappear after having
exceeded the elastic limit. The neglection of the bending
resistance has no influence on the calculation of the membrane tensile stresses in the web.

The edges of the web are rivetted to horizontal flanges and
vertical edge struts. It means a further disturbance of the
membrane stress condition that the edges are prevented from
buckling. But it can be demonstrated that this disturbance
is also negligibly small.

If the flanges and the edge struts were rigid and connected
with each other by pin joints, the direction of the wrinkles
would be exactly $45°$. The tensile stress is given by the
equilibrium condition, Fig.2.51b, as:

$$\sigma \cdot t \cdot (a \cdot \sin 45°) \cdot \sin 45° = P$$

$$\sigma = \frac{2P}{a \cdot t} \ .$$

Thus it is twice as high as the shear stress would be, when
the web could transfer shear stresses.

If one inserted between the two edge struts, additional
rigid struts, Fig.2.52, which are pin joined to the upper
and lower flange, then the direction of the largest tension

in the web, i.e. the direction of the wrinkles would not be changed; it would remain 45°. That means that the direction of the wrinkles is independent of the distance between the rigid edge profiles.

In reality, the flanges and struts are not rigid. Owing to the compressive stress in the vertical struts, the angle between the direction of the wrinkles and the flanges decreases to less than 45°. It generally amounts to $\alpha = 40°$ to 42°. Fig.2.53, upper frame.

2.5.2 The flanges

Due to the tension forces in the webs the flanges are bent in each panel between two vertical struts, Fig.2.53 lower frame. If the distance of the struts is chosen to be 1/6 to 1/2 of the height of beam, these bending moments have practically no deleterious influence on the strength of the flanges.

The struts have, on the contrary favorable influence on the flanges, since they give them narrow support and, hence, rise their buckling resistance. This resistance is increased furthermore, particularily in the case of open profiles, by the tension forces in the web, preventing the flanges to tilt.

2.5.3 The struts

The vertical struts support the two flanges against the tension forces in the web. Thus they undergo compression and are liable to buckling. Their buckling out of the plane of web is essentially aggravated by the tension stressed web, to which they are fastened. The calculation shows that their buckling load is 4 to 7 times as high as the Euler buckling

load of free struts having the same length. This high buckling strength is never utilized, because the elastic limit
would be exceeded long before the buckling load is attained.

2.5.4 Application

Finally it will be mentioned that an I-beam, which is shaped
as a tensile field, has not only low weight, but is also
very rigid.

The tension field is utilized mainly in aircraft structures.
In the case of aerofoil covering, for aerodynamic reasons,
buckling in normal flight condition should remain modest.
This can easily be attained, as only low torsion moments
exist in this state.

In Boeing 707, passengers can observe that, in a flight
attitude occurring temporarily during landing approach, the
surface of the aerofoil is traversed by a short-wave, straight
parallel postbuckling pattern. These wrinkles are not permanent: they will completely disappear after a few seconds.

Permanent wrinkles will not appear till the tensile stresses
in the web have nearly reached the elastic limit. The local
flexural stresses which occur owing to the formation of wrinkles have scarcely any influence on them.

2.6 Some remarks on optimization

2.6.1 Preliminary remarks

All papers till now discusssed in this chapter were written
by practical men, who deduced their scientific work from
questions that were brought to them owing to their function
as experts. Consequently, all these papers are aimed at

providing design data, immediately useful for the design engineer.

In opposition to this, van der Neut [17,18], whose papers will be discussed in the following, strives for fundamental findings. He deals with "The interaction of local buckling and column failure of imperfect thin-walled compression members". His theoretical investigations are carried through on the basis of assumed initial imperfections.

Van der Neut investigates rods, on which column buckling occurs without local buckling on the one hand, and local buckling without column buckling on the other hand, under loads of the same order of magnitude. He asks and answers the questions, whether for optimally designed structures the local buckling load must be smaller, equal or higher than the column buckling load. Torsional buckling is not being envisaged in these investigations.

A model has been taken which is representative of actual structures, as to the phenomena under consideration, and which is easy to handle analytically. This model is shown in Fig.2.54. It consists of two load carrying flanges with width b and thickness h , connected at distance 2·c by webs, which have no longitudinal stiffness. Concerning the buckling deformations the flanges are assumed to have, along their longitudinal edges, the boundary conditions

$$w = 0 \quad N_y = 0 \quad N_{xy} = 0 \quad M_y = 0$$

which are easy to take into account.

2.6.2 Perfect structures

In the unloaded condition the flanges are flat and the column axis is straight. Fig.2.55 shows the load shortening curve of

68

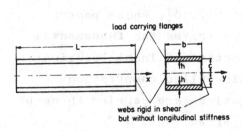

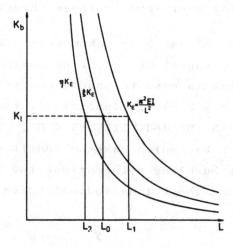

Fig.2.54 **Schematic view of the column.**

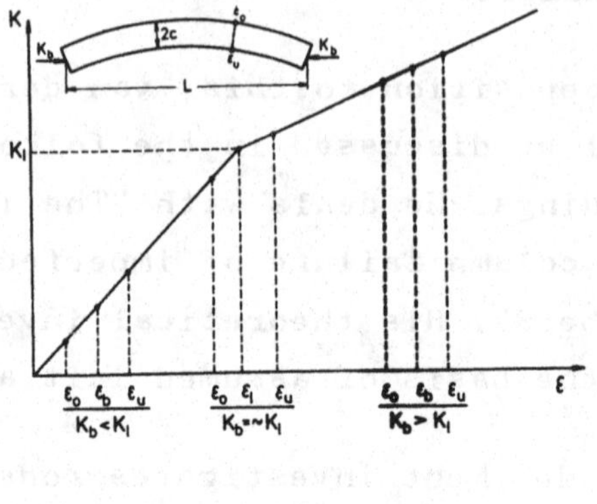

Fig.2.55 **Schematic load shortening curve.**

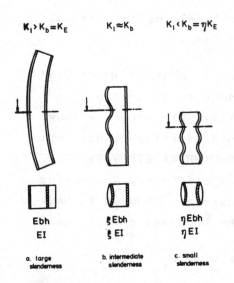

Fig.2.56 **Buckling behavior of columns with different slenderness ratios.**

Fig.2.57 **Column buckling loads as function of L.**

the box beam in the prebuckling and postbuckling region.
If both flanges are buckled, the slope of the curve is only
$\eta = 0.4083$ of the value prior to buckling.

For large slenderness of the column, Fig.2.56a, the buckling
load $K_b = K_E$ of the whole column lies below the buckling
load K_ℓ of the flanges. Hence, the stiffness of the flanges
is $E \cdot b \cdot h$ and the bending stiffness of the column is $E I$.

For small slenderness, Fig.2.56c, the buckling load K_ℓ of
the flanges lies below the buckling load K_b of the whole
column. When the whole column buckles, the flanges are already in the postbuckling state and their stiffness against
incremental compressive strain $\Delta\varepsilon$ is $\eta \cdot E \cdot b \cdot h$. Hence, the
bending stiffness under axial load is $\eta \cdot E \cdot I$ and the col-
umn buckling load is $K_b = \eta K_E$.

In between there is a transition state in which only one
flange is buckled. In this intermediate state the rigidity
of the column is

$$\xi \cdot E \cdot I = \frac{2\eta}{1+\eta} \cdot E \cdot I = 0.58\, E\, I$$

and the buckling load amounts to $K_b = \xi \cdot K_E$.

In Fig.2.57 the three curves K_E , $\eta \cdot K_E$ and $\xi \cdot K_E$ are
plotted versus the column length L . At first sight, nothing can be done with these curves, since we do not know
the transition from the curve for the unbuckled flanges K_E
to those for the buckled flanges $\eta \cdot K_E$.

From Fig.2.55 it can be deduced that an equilibrium state at
which only one flange is buckled, is possible only when the
load is approximately equal to the local buckling load K_ℓ .
Now this load will be plotted in the diagram 2.57. The hori-
zontal line $K_b = K_\ell$ intersects the curves $\xi \cdot K_E$, $\eta \cdot K_E$

and K_E in the points L_o , L_1 and L_2 .

Long columns $L > L_1$ buckle with the Euler load K_E and
the load carried by the structure remains constant in the
initial postbuckling region, as with every compact column,
cf. Fig.1.1 .

For columns in the length region between L_o and L_1 the
load can be increased up to the buckling load K_l of the
flanges only. Then the first flange buckles. Thereby the
load carrying capacity of the column decreases down to the
buckling load $\xi \cdot K_E$. That means that the equilibrium is
unstable and collapse will proceed explosively. Within this
region, imperfections have the effect, of reducing the buckling strength, to values below K_l .

For columns in the length region between L_2 and L_o the
load also increases up to the buckling load K_l of the
flanges. Then the first flange buckles. Thereby the load
carrying capacity will not decrease; theoretically the equilibrium state is stable. Practically it will be scarcely
possible, to load the column so cautiously, that only one
flange buckles and then no further increase of the load occurs. When both flanges buckle, the load carrying capacity
decreases down to the buckling load $\eta \cdot K_E$. In this length
region, the buckling load would be reduced by initial imperfections as in the zone between L_o and L_1 .

In the case of short columns $L < L_2$, the flanges buckle
before the buckling load of the whole column is reached.
When finally the whole column buckles the equilibrium state
is indifferent as in the case of the long column.

The diagram, Fig.2.57, was good for illustrating the fundamental relationships, but is has the disadvantages, that the

curves are too steep to allow an exact reading and that the
plot is not dimensionless. For this reason, van der Neut in-
troduces, for preparing his further statements, another type
of representation. Fig.2.58 shows the dimensionless failure
load K_b/K_l as a function of the dimensionless Euler load
K_E/K_l of the unbuckled column. The abscissa parameter,
contains only design data, and can be written in the form

$$\frac{K_E}{K_l} = \text{const.} \frac{1}{L^2} ,$$

which means it is nothing more than the square of the reciprocal column length.

We begin with the discussion of this diagram at the point
zero. For low values $1/L^2$, i.e. for long columns, the
column buckles as a whole, before the flanges buckle; therefore the dimensionless failure load K_b/K_l is equal to the
dimensionless Euler load K_E/K_l . This relationship remains
valid, with decreasing column length till the column has the
length L_1 , at which the buckling load of the flanges and
the Euler load of the whole column are equal. Now the failure of the column will be initiated by buckling of the
flanges. This failure load remains constant with further
decrease of the length, till the column has the length L_2 ,
at which the buckling load of the flanges coincides with the
Euler load of the column with buckled flanges. When the
column becomes shorter, its load carrying capacity increases
again, but at a lower rate than for the column with unbuckled
flanges.

2.6.3 Imperfect structures

In the previous section we simply stated that the column
in the length region between L_1 and L_2 is sensitive to

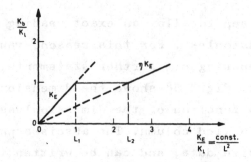

Fig.2.58 Column buckling loads as function of K_E/K_ι.

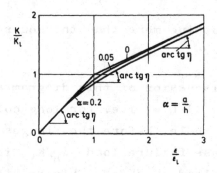

Fig.2.59 Load shortening curves of the imperfect structure.

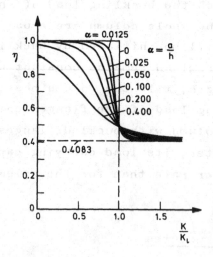

**Fig.2.60 Slope of the load shortening curves of the imper-
fect structure.**

initial imperfections. In the following this sensitivity
will be investigated numerically.

The behavior of the flanges is mainly governed by that
initial imperfection pattern, which corresponds to the
mode pertaining to the smalles load for local buckling.
The waviness was assumed to be given by

$$\frac{a}{h} = \alpha \cdot \cos\frac{\pi y}{b} \cdot \sin\frac{\pi x}{b} \; .$$

Fig.2.59 shows some load shortening curves which were calculated with this assumption for the initial imperfections.
For comparison the load shortening curve of the perfect
column, already known from Fig.2.55, is also plotted in the
diagram.

The slope of the normalized curves represents the ratio

$$\eta = \frac{d\left(\frac{K}{K_L}\right)}{d\left(\frac{\varepsilon}{\varepsilon_L}\right)} = \frac{\frac{dK}{d\varepsilon}}{\frac{K_L}{\varepsilon_L}} \; .$$

This is the reduction factor for the longitudinal stiffness
of the flanges against strain increment at its edges, caused
by bending of the column

$$\eta = \frac{\text{stiffness of the imperfect and/or buckled flange}}{\text{stiffness of the perfect, unbuckled flange}} \; .$$

For the perfect flange in the prebuckling region, there is
no stiffness reduction and, hence, $\eta = 1$; for the perfect
flange in the postbuckling region this reduction factor has
the well known value $\eta = 0.4083$. For the imperfect column
the slope differs, mainly in the region of the local buckling

load $K/K_l = 1$ from the slope of the perfect column. For
the loads $K/K_l < 1$ it is smaller and for $K/K_l > 1$ it is
higher.

Fig.2.60 shows the slope η as function of the normalized
load K/K_l for different initial imperfection amplitudes
$\alpha = a/h$. From this diagram it can be seen that, for imperfect columns, η decreases continuously. That is in opposition to the perfect column, for which η , at the point
$K/K_l = 1$, abruptly decreases from $\eta = 1$ to $\eta = 0.4083$.
As already stated, when discussing Fig.2.59, the slope is
lower for $K < K_l$ and is higher for $K > K_l$ than for the
perfect column.

As shown in Fig.2.56c for the perfect locally buckled column,
the relationship

$$K_b = \eta \cdot K_E$$

is valid. The same relationship holds for the imperfect
column. Consequently using the slopes η given in Fig.2.60
one can plot a diagram that represents the failure loads K_b ,
for different imperfection amplitudes α , as function of the
buckling loads K_E of the unbuckled column. This diagram is
shown in Fig.2.61 in dimensionless terms. You see the dashed
line of the perfect column which was already thoroughly discussed in Fig.2.58. The diagram further shows that the reduction of the buckling load by initial imperfections is
greatest in the region around $K_E/K_l = 1$, that the initial
imperfections have no influence on the buckling load at
$K_E/K_l = 2$ and that in the region $K_E/K_l > 2$ the failure
load is increased by the initial imperfections.

Postbuckling calculations must be performed using a nonlinear
shell theory and are, therefore, laborious. When calculating
Fig.2.59 to 2.61, van der Neut used an approximation which

becomes less accurate the larger the deformations are; therefore theoretical buckling loads for the region $K_E/K_L > 2$
should not be looked at too seriously.

But in the region $K_E/K_L = 1$, which is the most important
range the computed results are correct in their order of
magnitude. A comparison of these theoretical results with
experimental ones has not yet been performed, since the
theoreticians do not condescend to experimental work, whereas the practical men are downright occupied to verify their
own theoretical findings by experiments.

2.6.4 Stability in the initial postbuckling region

When discussing the perfect column, Fig.2.57, we stated that
the columns in the length region between L_2 and L_1 after
buckling are in an unstable equilibrium state, so that collapse would proceed explosively.

The stability in the initial postbuckling region was for the
first time investigated by Koiter [19]. In order to explain
the basic ideas of these investigations, we again point to
Fig.1.5, which presents the load shortening curves of perfect and imperfect structures. In the case of the plate,
the load shortening curve, after buckling, rises, i.e. the
equilibrium state in the initial postbuckling region is
stable as in the prebuckling region. For the cylinder, the
load shortening curve after buckling drops to lower loads
and lower shortenings: since the applied load or shortening
remains constant after buckling, the equilibrium state after
buckling is unstable.

The calculation of the postbuckling curves is, as already
mentioned, rather laborious. If one is only interested in

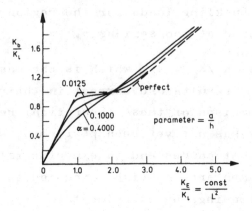

Fig.2.61 Buckling loads of columns with imperfect flanges.

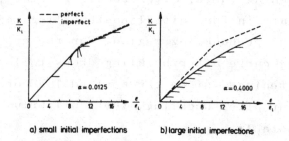

Fig.2.62 Stability in the initial postbuckling region.

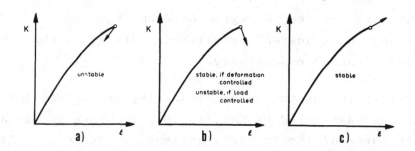

Fig.2.63 Load-shortening curves in the prebuckling region and tangents to the postbuckling curves at the bifurcation points.

the initial postbuckling region, one can use a much easier
calculation method which is valid only within the region of
very small deformations. One then obtains load shortening
curves as presented in Fig.2.62. Fig.2.62a shows a load
shortening curve in the initial postbuckling region, as it
is to be found with axially compressed cylinders: the equilibrium state is unstable, independent of, whether the cylinder is loaded by controlled shortening (ε = const) or
by controlled load (K = const) . Fig.2.62b shows a load
shortening curve, as it can be found at axially loaded cylinders with internal pressure: the equilibrium state is
stable in the case of controlled shortening (ε = const)
and is unstable in the case of controlled load (K = const).
Fig.2.62c shows an equilibrium state as it is to be found
in the case of the longitudinally compressed plate: it is
stable, independent of, whether the plate is loaded by controlled shortening or controlled load.

Van der Neut has calculated the load shortening curves in
the initial postbuckling region for the column with different imperfection amplitudes and different column lengths.
Fig.2.63a shows as an example the results of the calculation
for the small imperfection amplitude α = 0.0125 . You see,
as a thick line, the load shortening curve of the structure
before buckling as a column. Bifurcating from this line you
see load shortening curves for the initial postbuckling region pertaining to columns of different lengths. If the
column is long, it buckles at a small shortening, before the
flanges are buckled: then the load shortening curve in the
postbuckling region is a horizontal line. If the column is
shorter, within the critical length region between L_1 and
L_2 (see Fig.2.57) the load shortening curves point down-
ward. The equilibrium state in the initial postbuckling region is unstable, the structure will collapse. If the column

is still shorter L < L$_p$, it does not buckle before the
flanges are already buckled; now the equilibrium state in
the postbuckling region is stable, because the postbuckling
load of the flanges increases with increasing shortening.

For comparison, also the load shortening curve of the perfect structure, before column buckling, can be seen in the
figure. It scarcely differs from the load shortening curve
of the column with the small initial imperfection, so that
one can approximately assume that the postbuckling behavior
of the perfect and the slightly imperfect column are the same.

Fig.2.63b shows the corresponding postbuckling curve for the
column with the large imperfection amplitude $\alpha = 0.4$. The
column buckles under lower loads than the column with small
imperfections; but for compensation the equilibrium state
is stable after buckling, so that no drop of load and no
collapse occur.

This phenomenon that highly developed structures have a high
buckling load connected with a strong drop of load, whereas
comparable structures with lower buckling load buckle without or with low drop of load, is often found with thin-walled
shells. Therefore it seems reasonable to choose as the criterion for the load carrying capacity of thin-walled shells
not the buckling load but a suitable postbuckling load. This
will be discussed in more detail in the next chapter about
cylindrical shells.

But we are still dealing with van der Neut's investigation
of the postbuckling behavior of columns with box section.
Concluding, it should be mentioned, that the opinion, that
a buckling column constructed of several elements would be
optimally dimensioned, when the local buckling load and the

general buckling load are equal, is at least contestable;
the result would be that the column explosively collapses
when attaining the buckling load. Favourable postbuckling
behavior, i.e. low sensitivity to initial imperfections
and no drop of load after buckling, is guaranteed only,
when the local buckling load and the general buckling load
differ markedly.

Literature

[1] LAHDE, R.   Versuche zur Ermittlung der mittragen-
     WAGNER, H.   den Breite von verbeulten Blechen.
                        Luftfahrtforschung Bd.13 (1936),
                        S.214-218

[2] SCHNADEL, G.   Knickung von Schiffplatten.
                        Werft-Reederei-Hafen Bd.11 (1930)
                        H. 22/23

[3] MARGUERRE, K.   Die mittragende Breite der gedrückten
                        Platte.
                        Luftfahrtforschung Bd.14 (1937),
                        S.121-128

[4] COX, H.L.   Buckling of Thin Plates in Compression.
                        A.R.C. Rep. and Mem., No.1554, (1933)

[5] DAWSON, R.G.   Postbuckling of Geometrically Imper-
     WALKER, A.C.   fect Plates.
                        Proc. ASCE, J. of Structural Division
                        Vol.98 (1972), pp.75-93

[6] --   British Standards Institution, Speci-
                        fication for the Use of Cold Formed
                        Steel Sections in Building.
                        Addendum No.1 (1961) to B.S. 449

[7] CHILVER, A.H.   The Stability and Strength of Thin-
                        Walled Steel Struts.
                        The Engineer Vo.7 (1953), pp.180-183

80

[8]   WINTER, G.   Strength of Thin Steel Compression
                         Flanges.
                         Transactions ASCE, Vol.11 (1947),
                         pp.527-576

[9]   WINTER, G.   Performance of Thin Steel Compression
                         Flanges.
                         IABSE Preliminary Publications, Third
                         Congress, Liège (1948), p.137

[10] KLÖPPEL, K.   Die Berechnung der Traglast mittig und
      SCHUBERT, J.   außermittig gedrückter, dünnwandiger
                         Stützen mit kastenförmigem Querschnitt
                         im überkritischen Bereich.
                         Veröffentlichung des Inst.f.Statik u.
                         Stahlbau der TH Darmstadt (1971),H.13

[11] KLÖPPEL, K.   Die Traglast mittig und außermittig
      SCHMIED, R.   gedrückter,dünnwandiger Stützen mit
      SCHUBERT, J.   kastenförmigem Querschnitt im über-
                         kritischen Bereich unter Verwendung
                         der nichtlinearen Beultheorie.
                         Teil II: Experimentelle Untersuchungen,
                         Vergleich der experimentellen und theo-
                         retischen Ergebnisse.
                         Der Stahlbau Bd.38 (1969), S.73-83

[12] KLÖPPEL, K.   Ein Beitrag zur Ermittlung der Trag-
      UNGER, B.   fähigkeit dünnwandiger offener Pro-
                         file aus kaltgeformtem Stahl bei Be-
                         rücksichtigung des Zusammenwirkens
                         von Knicken, Drillen und Elementbeu-
                         len einschließlich der Berücksichti-
                         gung des überkritischen Beulverhaltens.
                         Veröffentlichung des Inst.f.Statik u.
                         Stahlbau der TH Darmstadt (1971), H.15

[13] --   Knickuntersuchung gemäß DIN 4114 /
                         Bl.1, 10.02 und 10.06

[14] ESSLINGER, M.   Kippen von Rahmenecken mit Rechteck-
                         querschnitt.
                         Der Stahlbau Bd.23 (1954), S.53-60

[15] KLÖPPEL, K.   Ein Beitrag zum Kipp-Problem von Rah-
      PROTTE, W.   menecken.
                         Der Stahlbau Bd.30 (1961), S.169-182

[16] WAGNER, H.   Über Konstruktions- und Berechnungs-
                        fragen des Blechbaues.

                        WGL-Jahrbuch (1928), S.113-125

[17] VAN DER NEUT,A. The Interaction of Local Buckling and
                        Column Failure of Thin-Walled Com-
                        pression Members.

                        Technological University of Delft,
                        Department of Aeronautical Engi-
                        neering, Report VTH 149 (1968)

[18] VAN DER NEUT,A. The Sensitivity of Thin-Walled Com-
                        pression Members to Column Axial
                        Imperfections.

                        Delft University of Technology,
                        Department of Aeronautical Engineering,
                        Report VTH 172 (1972) Aug.

[19] KOITER, W.T.   On the Stability of Elastic Equilibrium.

                        NASA TT F-10, 833

3. THIN-WALLED CIRCULAR CYLINDERS

3.1 Isotropic cylinders under external pressure

3.1.1 Experimental investigations

A thin ring loaded by radial compressive forces would
buckle with two waves around the circumference, Fig.3.1.
The same is valid for a thin-walled tube, the edges of
which can deform freely. These, however, are only mental
experiments.

The edges of actual test cylinders are cast into rigid
end plates and, hence, are restrained from deformation.
Cylinders with end plates buckle with two waves around
the circumference, only if they are sufficiently long.
Short cylinders buckle with larger circumferential wave
numbers.Fig.3.2 shows stable postbuckling patterns of
cylinders subjected to external hydrostatic pressure.
It is remarkable that all these patterns exhibit one wave
in axial direction and that the circumferential wave number increases with decreasing cylinder length.

Our test cylinders generally are manufactured from thin
Mylar foil. They have longitudinal seams, which are established by bonded one-sided butt strap joints. Mylar is a
plastic material with so high an elastic limit that the
buckling deformations vanish at unloading, and the cylinder may be used in more than one test.

In performing the tests, load-shortening curves are plotted.

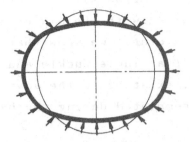

Fig.3.1 Buckled ring.

long cylinder   medium cylinder   short cylinder

Fig.3.2   Postbuckling patterns on cylinders of different
length.

Fig.3.3   Test device for buckling under external pressure.

Sometimes the buckling process with its unstable postbuckling patterns is recorded with the aid of high speed motion
pictures. From these pictures we know that buckling begins
with the appearance of a single buckle and that the stable
postbuckling pattern, existing in the advanced postbuckling
region is gradually completed during the buckling process.

In our tests with external pressure, the applied pressure
load is produced by reducing the pressure within the cylinder. The interior of the test cylinder is connected to
a large air chamber, Fig.3.3, so that the volume of the
enclosed air is sufficiently large, and hence, is not
greatly reduced by the snapping in of the buckles. In that
way the internal pressure and, consequently, the applied
differential pressure is kept nearly constant during the
buckling process. In the load-shortening diagram, Fig.3.4,
the snap-through line runs horizontally, i.e. at constant
load. After the snap-through the cylinder wall finds a new
stable equilibrium state with finite deflections. The load
can be increased further until, finally, the cylinder fails
by torsion.

Fig.3.5 shows postbuckling patterns of an isotropic cylinder subjected to external pressure at different levels.
The torsional pattern, in which the cylinder finally fails,
is caused by the axial component of the hydrostatic pressure load. We shall return to this phenomenon later when
extremely short cylinders under axial load are treated in
chapter 3.3.

In Fig.3.6 you see all postbuckling curves that could be
produced for this cylinder. The postbuckling curve, which
corresponds to the postbuckling pattern appearing spontaneously after buckling, was plotted down to its minimum.

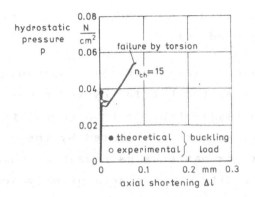

Fig.3.4 Experimental postbuckling curve.

a

b

Fig.3.5 Postbuckling patterns.

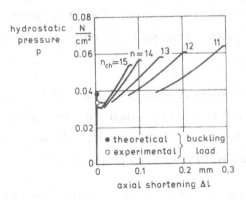

Fig.3.6 Experimental postbuckling diagram.

This was accomplished by gradually reducing the load until the buckles jumped out.

The snapping-in of the other load-shortening curves was manipulated, i.e. the buckled cylinder was stimulated by touching with finger tips to re-arrange its buckling pattern. Formerly we were fascinated by the production of as many postbuckling patterns as possible; nowadays we consider this activity as an amusement only. Nevertheless this amusement led to two fundamental findings:

- All the manipulated postbuckling patterns have smaller circumferential wave numbers than the pattern having snapped-in spontaneously.

- All the postbuckling loads, represented by the manipulated postbuckling curves, are higher than the minimum load of the postbuckling curve that has appeared spontaneously.

We denote the postbuckling pattern that snaps in spontaneously at buckling, as well as the corresponding postbuckling curve, and the smallest postbuckling load attained with this curve, as characteristic for the cylinder. It is important to keep in mind that the characteristic postbuckling load is the smallest load attained by the cylinder in the postbuckling region.

Fig.3.7 shows the measured buckling loads and the measured characteristic postbuckling loads of Mylar cylinders with dimensions $r = 100$ mm and $t = 0.254$ mm as function of the cylinder length. It can be noticed that

- the buckling loads and the characteristic postbuckling loads decrease with increasing cylinder length,

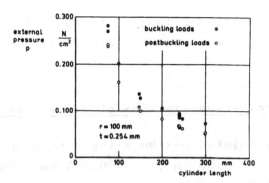

Fig.3.7 Experimental buckling and postbuckling loads.

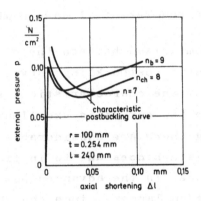

Fig.3.8 Selection of the characteristic postbuckling curve.

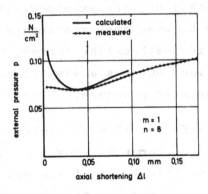

Fig.3.9 Theoretical and experimental characteristic postbuckling curve.

● for a given cylinder length, the buckling loads are
always higher than the characteristic postbuckling
loads.

3.1.2   Theoretical investigations of perfect cylinders

Fig.3.8 shows calculated postbuckling curves of a cylinder
with dimensions r = 100 mm , t = 0.254 mm and l = 240 mm
under external hydrostatic pressure [1]. It can be seen that
the curve, determined for n = 8   circumferential waves attains the lowest postbuckling load. Consequently, n = 8   is
the characteristic circumferential wave number.

Fig.3.9 shows the comparison between the calculated and the
measured characteristic postbuckling curves for this cylinder. In the region of the characteristic postbuckling load
the agreement of the two curves is excellent. At smaller
values of shortening there are considerable deviations since
in this range the disturbances due to initial imperfections
are of great influence. In the advanced postbuckling region
the deviations are even larger, since the shell theory which
was used is valid only in the elastic region and for small
deflections, whilst the deformations have grown so large
that both these limitations were exceeded. It may be noticed
that the correlation between the experimental and the theoretical characteristic postbuckling curves, presented in
Fig.3.9 is no exception, but was always found for cylinders
under external hydrostatic pressure.

The theoretical circumferential wave numbers of the buckling
patterns are somewhat higher than those of the characteristic
postbuckling patterns. This theoretical result can not be
checked by experiments, since the regular buckling pattern
does not appear. The high speed motion pictures demonstrate

only that the buckles snap in one after the other until the stable postbuckling pattern is completed.

Fig.3.10 shows calculated and measured characteristic circumferential wave numbers plotted against the cylinder length. It can be seen that the agreement is good and that the wave numbers decrease with increasing cylinder length.

Fig.3.11 shows the experimental buckling loads, the characteristic postbuckling loads calculated for the perfect cylinder and the experimental characteristic postbuckling loads as functions of the cylinder length. It can be seen that

- the experimental and theoretical postbuckling loads are in good agreement,

- the buckling loads and the characteristic postbuckling loads decrease with increasing cylinder length,

- the experimental buckling loads lie in a scatter region, the lower limit of which is given by the theoretical characteristic postbuckling loads of the perfect cylinder. The upper limit of course is given by the theoretical buckling loads.

This scatter region of the buckling loads of imperfect cylinders between limits which could be calculated for the perfect cylinder is exhibited in dimensionless terms in Fig.3.12. The load parameter

$$\bar{p} = \frac{p}{E} \cdot \left(\frac{r}{t}\right)^2 \cdot \sqrt{3\left(1-\nu^2\right)}$$

and the length parameter

$$\bar{l} = \frac{l}{\sqrt{r \cdot t}} \cdot \sqrt[4]{3\left(1-\nu^2\right)}$$

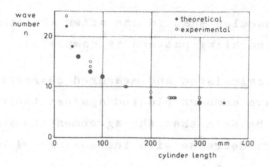

Fig.3.10 Circumferential wave numbers.

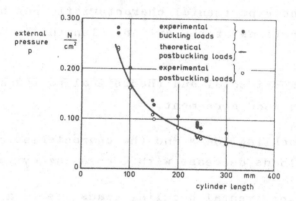

Fig.3.11 Theoretical postbuckling loads and experimental buckling and postbuckling loads.

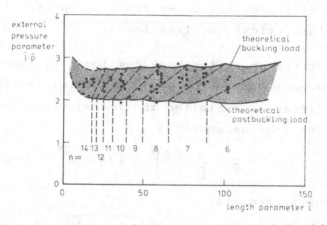

Fig.3.12 Scatter region of the experimental buckling loads.

are based on similarity principles [3]. As ordinate the
product $\bar{\iota} \cdot \bar{p}$ was chosen, because this quantity, for a
wide region of the parameters $\bar{\iota}$ is independent of the
cylinder length. The circumferential wave numbers n indicated in the figure are valid for the radius-to wall-
thickness ratio r/t = 384 only. The envelopes to these
curves would fix the limits of the scatter region for all
isotropic cylinders subjected to external hydrostatic pressure; we have checked this statement by a large number of
buckling tests.

It may be mentioned that the nearly horizontal course of
the limiting curves in Fig.3.12 does not go on without
bounds since the buckling pressure of the long cylinder is
limited by the buckling pressure of a ring. The smallest
circumferential wave number is n = 2 . Postbuckling investigations for extremely long cylinders have not been
performed either experimentally or theoretically so far.

3.1.3 Theoretical investigations of imperfect cylinders

The limitation of the experimental buckling loads in the
way demonstrated in Fig.3.12 may seem somewhat strange, and
someone might ask, whether it is not simpler to determine
the buckling load of the imperfect, hydrostatically loaded,
cylinder as function of measured or reasonably assumed
imperfections, as it is done with columns and plates.

This question would be incorrect, since it is not true that
the buckling loads of columns and plates can be calculated
as functions of the imperfections. In order to prepare the
proof for this assertion, the concept "buckling load" shall
be defined clearly in the following: We once more look at
the fundamental picture, Fig.1.5, and there we notice that

for all perfect structures bifurcation points of the load-
shortening curves exist. At a bifurcation point, two different infinitely neighbouring equilibrium states are possible
at the same load. This bifurcation load is the buckling
load of the perfect structure. The equilibrium states corresponding to the deviating branches of the load-shortening
curves are different for the three types of structures with
respect to their stability. For the column they are neutral,
for the plate stable and for the cylinder unstable.

In the case of imperfect structures there are also bifurcation points in fact always then if the prebuckling pattern is orthogonal to the buckling pattern. The classical
example for this type of bifurcation is the cylinder under
axial compression with an axisymmetric prebuckling pattern,
Fig.3.13, left hand side, that becomes unstable with an
asymmetric buckling pattern, Fig.3.13, right hand side. But
this kind of buckling loads shall not be considered here.

The load-shortening curves of imperfect structures, plotted
in Fig.1.5, exhibit no bifurcation of equilibrium, but it
is assumed that the prebuckling deformation continues to
grow under increasing loads till the load-carrying capacity
of the structure is exhausted. In the case of beams and
plates under these assumptions, there is no instability of
equilibrium states in the region of finite elastic deformations: in beams and plates the ultimate load is defined as
the load at which the yield limit of the material is attained. Strictly speaking one must not talk of "imperfect
plates in the postbuckling region", but should correctly
speak of "imperfect plates in the hypercritical region".
The term "hypercritical" means a loading that is above the
buckling load of the perfect structure, and has been attained with continuously growing deformations, i.e. without

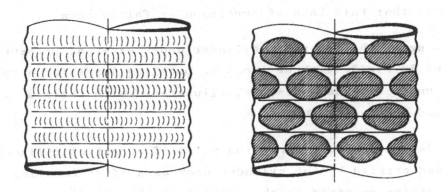

Fig.3.13   Axisymmetric imperfections and buckling pattern.

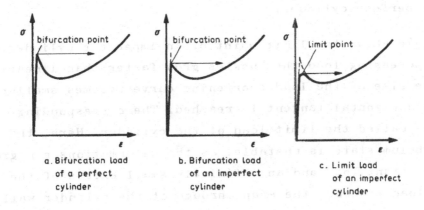

a. Bifurcation load
   of a perfect
   cylinder

b. Bifurcation load
   of an imperfect
   cylinder

c. Limit load
   of an imperfect
   cylinder

Fig.3.14   Three types of buckling loads.

a sudden buckling process. Consequently the correct title
of this book should have read "hypercritical and postbuckling behavior of structures". Our excuse for the incorrect
title is that this lack of precision is fairly common.

Returning to the imperfect cylinder under external pressure
we find in Fig.3.14 three types of buckling loads for a cylinder under external pressure, illustrated with the aid of
the load-shortening curves.

Fig.3.14a shows the bifurcation point of a perfect cylinder.
The load carried by the cylinder decreases after buckling.
The equilibrium state in the initial postbuckling zone is
unstable. However, the cylinder finds a new stable equilibrium configuration with large deformation in the deep
postbuckling region.

Fig.3.14b shows the bifurcation point of an imperfect cylinder. The postbuckling behavior is exactly similar to that
of the perfect cylinder.

Fig.3.14c shows the limit point of an imperfect cylinder.
With increasing load the buckles grow faster than linearly
and the rise of the load shortening curve becomes smaller
till a horizontal tangent is reached. The corresponding
load is called the limit load of the cylinder. Here, the
equilibrium state is unstable, as the deformations can grow
with constant load and an infinitely small excess of the
limit load produces the snap-through of the cylinder wall,
to a new stable equilibrium configuration with large deformation in the advanced postbuckling region.

Summarizing, we state that in the case of thin-walled cylinders under hydrostatic pressure there are two types of

buckling loads: the bifurcation load of perfect or imperfect cylinders and the limit load of imperfect cylinders.

For thin-walled shells of simple shape, limit loads can be calculated using initial imperfection patterns which are affine to the buckling patterns. The measured imperfection pattern is irregular, but for cylinders under external pressure the actually measured imperfection pattern can be replaced by a simplified pattern, which permits a relatively inexpensive computation of the buckling load of the imperfect cylinder. If this simplification is performed reasonably, this buckling load is equal to that, which would be obtained by the use of the exactly measured initial imperfection pattern.

Now, you will understand my answer to the question that was raised at the beginning of the section, whether the buckling load of a cylinder under external pressure can be computed as a function of the initial imperfections, as customarily done with columns or plates:

- First, the question is not correct, since the ultimate load of imperfect plates and columns is determined by attaining the yield stress, rather than by buckling of the structure.

- Second, yes, it is possible to compute the buckling load of cylinders under hydrostatic pressure as function of the initial imperfections.

Before this computation can be attacked, two conditions must be fulfilled. One must know:

- the pattern with which the imperfect cylinder will buckle,

● the amplitude of the simplified initial imperfection
pattern, which is to be used in the analysis.

Although one cannot exactly predict the buckling pattern,
yet one can, on the basis of test observations, deduce the
bounds of the possible patterns with some thought. Cylinders which are slightly imperfect will buckle with the
same pattern as the perfect cylinder, while strongly imperfect ones will buckle with the characteristic postbuckling
pattern, or not at all.

These statements will be elaborated a bit further. In
Fig.3.15a you see load-shortening curves, corresponding
to the buckling pattern of the perfect cylinder $n_b = 9$;
the uppermost one is the postbuckling curve of the perfect
cylinder, and the two others are load-shortening curves of
imperfect cylinders. In the deep postbuckling region the
three postbuckling curves merge, since here the postbuckling deformations have become so large that the initial
imperfections are negligible in comparison. The intermediate load shortening curve corresponds to small imperfections and exhibits a limit point at which the equilibrium becomes unstable and the cylinder snaps through. The
lower curve corresponds to large imperfections and does
not result in a snap-through because the minimum of the
perfect postbuckling curve lies so far above that the imperfect curve can go under it.

Fig.3.15b shows load-shortening curves corresponding to the
characteristic postbuckling pattern $n_{ch} = 8$. For comparison the postbuckling curve $n_b = 9$ is also indicated. The
buckling load of the perfect cylinder is the point, where
the postbuckling curve $n_b = 9$ bifurcates from the prebuckling line. It is lower than the point, where the characteristic postbuckling curve $n_{ch} = 8$ bifurcates. Besides

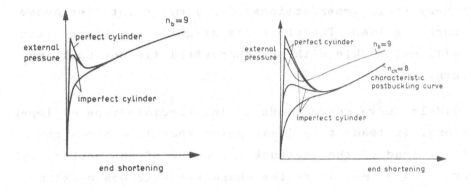

a) $n_b = 9$ b) $n_{ch} = 8$

Fig.3.15   Schematic load-shortening curves of imperfect cylinders .

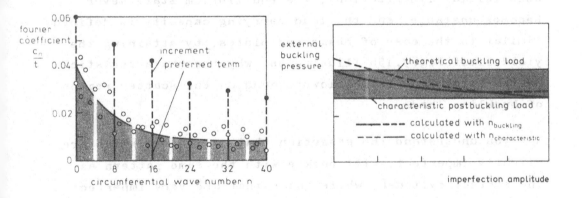

Theoretical buckling loads of imperfect

Fig.3.16 Simplification of the measured initial imperfection pattern .

Fig.3.17 Determination of the buckling load of an imperfect cylinder .

the two postbuckling curves of the perfect cylinder you
can find in the figure three postbuckling curves of imperfect cylinders. The uppermost corresponds to a cylinder
with very small imperfections; its limit point lies above
the buckling load. Therefore the slightly imperfect cylinder will not buckle with the characteristic postbuckling
pattern.

The middle curve corresponds to intermediate type of imperfections. It leads to a limit point that lies below the
smallest load of the postbuckling curve of the perfect cylinder $n_b = 9$ and above the characteristic postbuckling
load. This theoretical limit point may be attained in tests.

The lowest postbuckling curve corresponds to large imperfections; it has no limit point but goes below the characteristic postbuckling load in the deep postbuckling region.
This postbuckling curve is also real. If the cylinder has
deep initial imperfections, the equilibrium state never
becomes unstable, and the load carrying capacity is defined,
similar to the case of beams and plates, by attaining the
yield limit. Fig. 3.15b illustrates, why the characteristic
postbuckling load is the lower bound on the scatter region
of buckling loads.

Now you understand the assertion that cylinders, which are
slightly imperfect, will buckle with the same pattern as
the perfect cylinder, while those that are very imperfect,
will buckle with either the characteristic postbuckling
pattern, or not at all.

We come to the second factor governing the determination of
the buckling load of an imperfect cylinder: We should have
a reasonable measure of the amplitude of the initial imper-

fection pattern. There are two basically different possi-
bilities for obtaining it:

The first method, for finding an initial imperfection amplitude, consists of starting from a measured buckling load,
assuming a reasonable buckling pattern and working backwards
to determine how large the initial imperfection amplitude
of this pattern should have been. This value of prebuckling
amplitude may be carried over to other cylinders with the
help of similarity principles. This is a very useful method
as long as one deals with cylinders of similar type and similar grade. We recall that the researchers in plate theory,
whose formulas for the load carrying capacity are given in
chapter 2, have proceeded exactly in the same fashion.

The second method, for finding an initial imperfection amplitude, consists of measuring, the deviation of the cylinder from the ideal shape, and then simplifying the initial
imperfection pattern. These types of measurements have been
carried out in several research centers. At the DFVLR in
Braunschweig we measure with a non-contacting capacity pick-
up to an accuracy of 1/1000 mm. The measured initial imperfection pattern is represented in a double fourier-series.

$$f_{measured} = \sum_m \sum_n \left(c_{mns} \cdot \sin \frac{ny}{r} + c_{mnc} \cdot \cos \frac{ny}{r} \right) \cdot \sin \frac{m\pi x}{l} .$$

Since cylinders, subjected to external pressure, buckle
with one-half wave over the cylinder length, all the fourier
terms that have an axial wave number m greater than one,
are set to zero, as they are orthogonal to the buckling pattern. Now there remain for each value of the circumferential
wave number n a cosine term and a sine term. We calculate
from it the resulting amplitude without worrying about the
phase shift y_n.

$$f_{reduced} = \sum_n c_n \cdot \cos \frac{n(y+y_n)}{r} \cdot \sin \frac{\pi x}{l} \, .$$

Our double-fourier series is now so reduced that it contains only terms with m = 1 and that for each value of n
there is only one term. Of these terms those with the circumferential wave numbers, coinciding with that of the buckling pattern or an integer multiple of it, are taken with
their actual values. They are the preferred terms. The remainder is taken into account by an increment to each of
the preferred terms, Fig.3.16. We have given a great deal
of thought to the computation of this increment (that is an
euphemism for saying that it is arbitrary) but I do not wish
to go into any further now.

If one were, to simply add these selected fourier coefficients, one obtains a theoretical depth of buckle which is
larger than the largest measured buckle-depth, as the phase
shift has been neglected. In order, therefore, to compensate
for it, we introduce a correction factor K .

$$f_{simplified} = K \cdot \sum_j c_j \cdot \cos \frac{jn_0 y}{r} \cdot \sin \frac{\pi x}{l} \, .$$

Since one does not know the buckling pattern exactly, one
reduces the imperfection pattern and computes the limit load,
based on the buckling pattern of the perfect cylinder, on
the characteristic postbuckling pattern and, eventually, on
the set of circumferential wave numbers that lie between the
two. The smallest limit load resulting from this procedure
is taken as the buckling load for the imperfect cylinder.
This calculated buckling load was found in our investigations, to be about 5 % off from the measured buckling load,
Fig.3.17.

The simplified representation of the measured initial imperfections has sufficient manoeuvrability that for one
cylinder agreement between the theoretical and the experimental buckling load could be obtained. The process cannot
be considered reliable before correct buckling loads have
been found, for a number of different cylinders with the
same assumptions.

Whether the expense of computing the buckling load as a
function of the measured or assumed initial imperfections
is worthwhile or whether one can more effectively compute
the lower bound of the scatter zone of buckling loads, that
is the characteristic postbuckling load, Fig.3.12, remains
to be seen.

3.2   Isotropic cylinder under axial load

3.2.1   Preliminary remarks

Now we turn to isotropic cylinders under axial loads. The
history of the investigations on the postbuckling behavior
of thin-walled cylindrical shells under axial load can be
compared to that of the days of Gold Rush in Wild West,
with the seekers of El Dorado finding some small prize
amidst much disappointment but with unshakeable hope.

The experimental buckling loads of thin-walled cylinders
under axial load are independent of the cylinder length
with the exception of extremely short or extremely long
cylinders. The deviation of the experimental results from
the theoretically predicted buckling loads of perfect cylinders is considerably greater in the case of axial compression than in the case of external pressure. For extremely thin-walled cylinders the test results may be only 20 %
of the theoretical load as shown in Fig.3.18.

Recognizing that this discrepancy of theoretical and experimental results is mainly due to the postbuckling behavior of thin-walled shells, v.Kármán and Tsien [4] in an
epochmaking paper, came up with a theoretical postbuckling
investigation based on diamond shaped postbuckling patterns,
Fig.3.19. In this analysis the buckling patterns were taken
as repeating endlessly without the necessity of strictly
enforcing boundary conditions. For the postbuckling behavior one must use a non-linear shell theory. The deflection
pattern and the calculations were rough approximations.
There was no other possibility at that time, since these
pioneers had no modern digital computers at their disposal.
The principal results of this study was that the theoretical postbuckling loads were lower than the theoretical buckling loads. This pioneering publication entailed a flood of
postbuckling investigations based on theoretical and experimental methods.

3.2.2 Experimental investigations of stable equilibrium states

Attempts have been made to decrease the discrepancy between
the experimental and the theoretical buckling loads by using
carefully manufactured test cylinders that were nearly free
of initial imperfections. These cylinders were made out of
Mylar as well as metal. In exceptional cases seamless cylinders were manufactured by electroforming out of copper or
nickel or by centrifugal casting out of synthetic resin.

The test cylinders were placed between parallelly guided
plates and loaded under controlled deformation. Fig.3.20
shows the test device used at DFVLR Braunschweig. In this
equipment axial loads up to $40 \cdot 10^4$ N can be applied. The
upper end plate, before the test, is adjusted to the cylinder

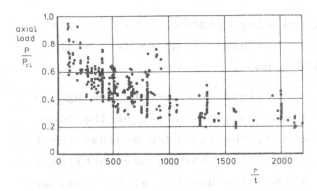

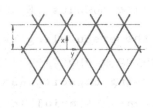

Fig.3.18 Experimental buckling loads.   **Fig.3.19 Diamond shaped postbuckling pattern.**

Fig.3.20 Loading device.

length and fixed in this position. The lower plate is
moved upward at a velocity between 0.01 and 1.0 mm/min.

Fig.3.21 shows a load-shortening diagram, obtained from a
typical test. The cylinder was made of Mylar foil with
Young's modulus $E = 5500$ N/mm^2 .

In the prebuckling region the load and the shortening increased proportionally. When the upper limit of the load
carrying capacity was reached, the cylinder avoided further
increase of axial load by buckling. During the buckling process the end-shortening remained constant, as the test apparatus is a deformation-controlled one. The cylinder wall
reached a new stable equilibrium configuration at a load
far below the buckling load and with finitely large, visible
deformations. If then, the shortening was increased further,
the load carrying capacity increased till a secondary buckling load was reached and so on.

The symbols "2,15" on the first branch of the postbuckling
curve stand respectively for the number of buckles, seen
longitudinally and around the circumference. After the first
secondary buckling the postbuckling pattern "2,14" appeared;
with further shortening the pattern progresses sequentially
to one denoted by "2,10".

With increasing shortening the postbuckling loads and the
number of buckles in the circumferential direction decrease.
After the pattern denoted by "2,10", an irregular pattern
emerged which eventually changed to a pattern denoted by
"1,6". Finally we stopped further end-shortening in order
to prevent the cylinder from severe damage. In retracing
the load-shortening curve we cannot follow the path that was
traced during the loading phase, as the bifurcation points
are not the same during loading and unloading. Hence,

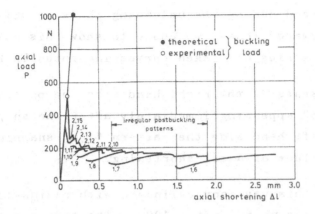

Fig.3.21 Load-shortening curves r = 100 mm ,
t = 0.19 mm , 1 = 100 mm .

one tier n = 6   two tier n = 15

Fig.3.22 Postbuckling patterns .

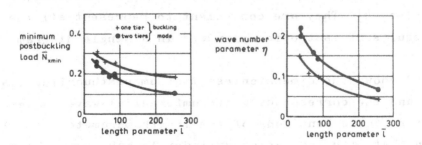

Fig.3.23 Minimum postbuckling loads and corresponding
circular wave numbers .

any curve traced during the unloading phase is different from the "forward" phase. We chose to show this unloading phase by the series of broken curves indicated in Fig.3.21.

Fig.3.22 presents at the right hand side the postbuckling pattern "2,15" appearing immediately after the snap-through, and at the left hand side the pattern "1,6" snapping in at very large values of axial shortening.

We tested in this way many cylinders with radius-to-wall-thickness ratios up to $r/t = 1000$. The test results have been presented in terms of the following dimensionless parameters:

the length parameter

$$\bar{l} = \frac{l}{\sqrt{r \cdot t}} \cdot \sqrt[4]{3(1-\nu^2)} \,,$$

the load parameter

$$\bar{N} = -N_x \cdot \frac{r}{E \cdot t^2} \cdot \sqrt{3(1-\nu^2)}$$

and the circumferential wave number parameter

$$\eta = n^2 \cdot \frac{t}{r} \cdot \frac{1}{\sqrt{3(1-\nu^2)}} \,.$$

The parameters $\bar{l}$, $\bar{N}$ and η are based on similarity principles [4]. They are convenient to represent all the test results of isotropic cylinders in a single diagram.

Fig.3.23 shows the dimensionless minimum postbuckling loads N_{xmin} and the corresponding circumferential wave number parameter η as functions of the length parameter $\bar{l}$. It is seen that with increasing cylinder length, both the postbuckling loads and the wave numbers decrease.

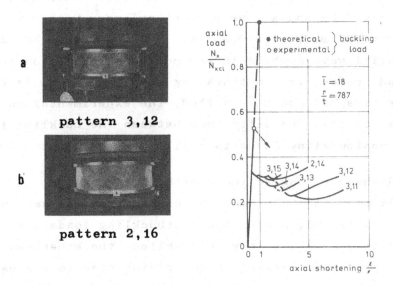

Fig.3.24 Postbuckling behavior of a short cylinder .

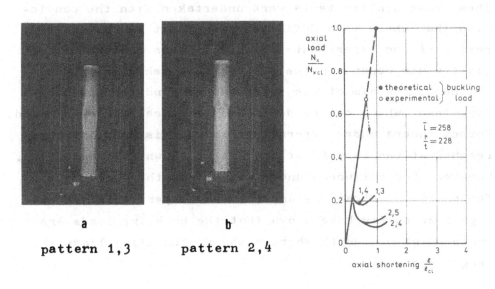

Fig.3.25 Postbuckling behavior of a long cylinder .

We would like to illustrate this rule by two extreme examples. Fig.3.24 shows a very short cylinder with dimensions r = 200 mm , t = 0.254 mm , ℓ = 100 mm . The circumferential wave numbers in the postbuckling region are
large and, hence, the postbuckling loads are high. As the
cylinder is short and thin-walled, the experimental buckling load is low. The difference between the buckling loads
and the postbuckling loads is small.

Fig.3.25 shows a long cylinder with dimensions r = 58 mm ,
t = 0.254 mm , ℓ = 770 mm . The circumferential wave numbers are small and, hence, the postbuckling loads are low.
As the cylinder is not very thin-walled, the experimental
buckling load is relatively high, giving rise to a great
difference between the buckling loads and the postbuckling
loads.

These postbuckling tests were undertaken with the convic-
tion that there is a definite lower limit to the scatter
region of the experimental buckling loads and that this
limit would depend on the postbuckling behavior of the
cylinder. The aim of the tests was to find the postbuckling load, which can be defined as the desired lower bound.
For cylinders under external pressure this load had been
reached at the very first attempt as shown in chapter 3.1.
However, for cylinders under axial load these attempts
failed in the first instance, as the experimental investigations so far have shown that the buckling loads are
independent of length whereas the postbuckling loads are
length-dependent.

Before concluding these hopeless reflections, we would like
to show another diagram, Fig.3.26, where the lowest postbuckling loads of cylinders with different lengths are

plotted versus the wave number parameter η . The numbers
on the curves indicate the length parameter l . One sees
here, that the postbuckling loads for one- and two-tier
postbuckling patterns follow one curve each, and that the
loads and circumferential wave numbers are the lowest ones
for the longest cylinder. This diagram was drawn and brought
to the first author by a colleague who is a matter-of-fact
and phlegmatic scientist, with the remark that this curve
would be heard around the world. Since that time, we call
this diagram the "curve that will be heard around the world".

However, this proud hope was not fulfilled; for the principle illustrated by this diagram shows that we have got to
a cul-de-sac. Investigations on the postbuckling behavior
as described above can not lead to a definition of a lower
limit on the scatter band of the buckling loads as the postbuckling loads are length dependent, while the buckling
loads are length independent.

3.2.3 Theoretical investigations of stable equilibrium states

The theorists have, in the meantime, laboured hard to improve upon v.Kármán's postbuckling calculation, in order to
find the lowest postbuckling load. This load was to be the
lower limit of the scatter region of the experimental buckling loads. In v.Kármán's approach, the postbuckling pattern was represented by a series. The coefficients of this
series were determined from the condition that the potential
energy must be a minimum.

The researchers, who followed v.Kármán in the first instance,
considered not only the coefficients of the series terms,
but also the wave length of the buckles as variable parameters

of the minimization problem. They obtained as a result of
their calculation a single postbuckling curve, Fig.3.27a.
As this curve was based on dimensionless parameters, it was
valid for all isotropic cylinders [6]. It was independent
of cylinder length as it was based on the diamond-shaped
postbuckling pattern, Fig.3.19, extending boundlessly along
the cylinder.

The lowest postbuckling load, computed on this basis, was
presumed to be the sought-after lower limit of the test-
scatter zone. More and more terms were taken in the series
expression and as lower and lower values of the postbuckling
loads were obtained, there was much feeling of complacency.

Fig.3.27b presents a typical diagram from a publication of
those times [6], showing the progress in the computation
of the postbuckling load. The lowest curve, according to
the proud claim of the author, agreed very well with test
results. The cited test results, naturally, were postbuckling loads of cylinders with finite length, hence, the comparison is not meaningful. The theoretical paper entirely
ignored the length-dependence of the experimental postbuckling loads.

This feeling of complacency in the progress toward a lower
limit of postbuckling loads came to an abrupt end, when
Hoff [7] in a lecture presented in Paris in 1964 somewhat
doubtfully made the remark: "It appears that these calculations based on many terms of series will ultimately give us
a postbuckling load of zero".

Hoff's results are not easy to explain, as he had found,
using dimensionless parameters, that simultaneously with
the load parameters $\bar{N}_x$ the wave parameter η also tended

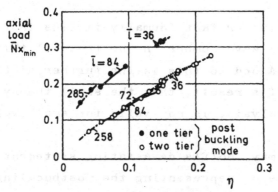

Fig.3.26   Minimum postbuckling loads as function of wave
           parameter .

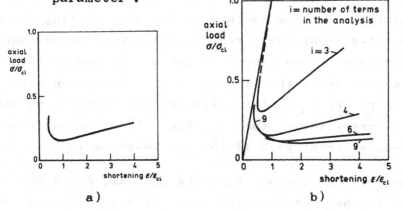

a)   b)

Fig.3.27   Development of the single postbuckling curve .

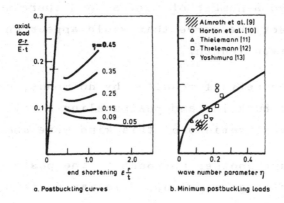

a. Postbuckling curves   b. Minimum postbuckling loads

Fig.3.28   Postbuckling curves calculated with fixed values
           of wave number n [8] .

to zero. This latter fact formally implies that either the circumferential wave number n or the wall-thickness t goes to zero. Added to this is the further difficulty, in explaining Hoff's result, that the shell theory, he had used in his analysis, is not valid for small values of n .

Be that as it may, Hoff's devastating statement put an end to the practice of representing the postbuckling behavior of isotropic cylinders by a single postbuckling curve.

One explanation of the failure of the postbuckling computations has been that the circumferential wave number had been treated as a continuous variable and, hence, the periodicity condition in the circumferential direction was violated. Therefore the later postbuckling calculations made use of discrete values for n . Fig.3.28 shows an example taken from a report by Hoff [8] published in 1965.

When one regarded the postbuckling diagram, Fig.3.28a, one rejoiced in the fact that the postbuckling loads are not zero. Comparison with experimental values showed that the minimum postbuckling loads were in reasonable agreement for equal wave number parameters η , Fig.3.28b. However, the theory produced a number of curves, and there was no criterion to select the curve that would appear on an actual cylinder of given length.

Since these theoretical results, by no means, could be used to predict the buckling and postbuckling behavior of actual cylinders, investigations of this kind were abandoned.

At this juncture another approach to the postbuckling calculation began to gain ground. This was Koiters method [14] to investigate the stability in the initial postbuckling region with a perturbation scheme. In section 2.6, an

example of a slender rod was considered, and there it was
shown that the sensitivity to initial imperfections depends
on whether the load-deflection curve rises or falls in the
initial postbuckling region, Fig.2.61. Similar calculations
using Koiter's perturbation method for axially compressed
cylinders confirmed the experimental finding that isotropic
cylinders under axial compression are extremely sensitive
to initial imperfections.

However, further investigations revealed that the behavior
of the shell in the initial postbuckling region is not a reliable criterion for its initial imperfection sensitivity.
Kempner [15] showed at the 12th International Congress of
Applied Mechanics at Stanford University in 1968 that on an
elliptical cylinder under axial loading, immediately beyond
the bifurcation point the postbuckling curve drops a bit and
then again rises steeply. This signifies physically that the
weakly curved regions of the ellipse buckle first, and then
a re-distribution of the load occurs so that in the advanced
postbuckling region the load is taken up mainly by the strongly curved portions of the ellipse, which permits the load
shortening curve to rise again. Fig.3.29 shows an example
by Kempner [15]. One can conclude from it that even though
there is a slight drop in the load shortening curve in the
initial postbuckling region, the elliptical cylinder is not
sensitive to imperfections. Hence, the initial postbuckling
behavior does not yield a reliable criterion for the sensitivity to initial imperfections.

Koiter's theory withstood the attack and is being used in
numerous publications. It has the advantage of requiring
very little computational time, and yet, yielding correct
postbuckling curves in the initial postbuckling region.How-
ever, the extension of this calculation method into the advance postbuckling region is not possible.

At the DFVLR Braunschweig we interpreted Hoff's remark
about the postbuckling load based on v.Kármán's pattern
tending to zero, to mean that reasonable postbuckling calculations must take into account the finite length of the
cylinder, as otherwise one calculates the stability of an
infinitely long slender column, which clearly leads to
zero buckling load [16]. How far this interpretation was
correct may be left aside. Its important effect was that
we now began to calculate the load-shortening curves with
the length of the cylinder and the actual test boundary
conditions taken into account. In our tests the cylinder
end sections are cast into rigid bulkheads and the test
specimens are compressed between parallelly guided plates.

In the calculations we used a series expansion with 50
terms. Needless to say, the dimensions of the cylinder
were those of the test specimens for which we had the experimental postbuckling curves. It might be added that this
is in contrast to the usual postbuckling investigations
where the comparisons are made with somewhat arbitrarily
selected postbuckling curves. Fig.3.30 shows the comparison
of a theoretical and an experimental postbuckling curve of
a Mylar cylinder with the dimensions r = 100 mm, t = 0.254 mm
and l = 300 mm . It can be recognized that for equal circumferential wave numbers the results agree within the in-
accuracy of experiments. But the theoretical postbuckling
investigations have resulted in more than mere agreement of
arbitrarily chosen postbuckling curves. The comparison of
theoretical postbuckling diagrams with test results yielded
the criterion to distinguish the postbuckling curves that
will be found in tests: The curve attaining the smallest
postbuckling shortening in the theoretical postbuckling
diagram, Fig.3.31, will appear immediately after the snap-
through.

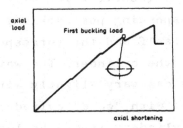

Fig.3.29 Postbuckling curve of an elliptical cylinder .

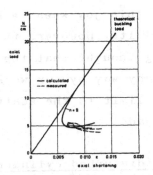

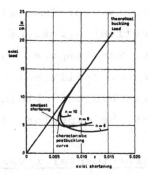

Fig.3.30 Theoretical and experimental load shortening curves .

Fig.3.31 Selection of the characteristic postbuckling curve .

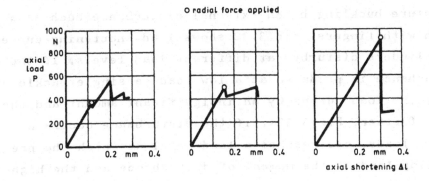

a) low level   b) medium level   c) high level

Fig.3.32 Load shortening curves of disturbed cylinders.

We named the postbuckling curve which attains the smallest
shortening, the corresponding postbuckling pattern, the
smallest shortening itself and the corresponding load as
the characteristic of the cylinder. The wave number observed in experiments can vary slightly with the quality
of the test specimens; with "good" cylinders it can be
higher and with bad cylinders it can be lower.

It is a rare moment of rejoicing in the life of a research
worker, when he finds that his theoretical predictions are
confirmed by experiments; and we felt this joy when we had
this excellent agreement with test results. However, this
rejoicing was tempered by the sobering thought that the
designer had really not benefited in any way by our work,
as we still had not obtained the lower limit to the scatter
region of the experimental buckling loads.

3.2.4 Experimental investigations of disturbed cylinders

Now we hoped to find the desired lower limit of the scatter
region of the buckling loads using a rather primitive experimental approach; we stimulated the loaded cylinder to
premature buckling by an external disturbance such as a
touch with fingers. Fig.3.32 shows load-shortening curves
of cylinders disturbed at different load levels. If the
disturbance is produced at a low load, a single buckle occurs; the load drops by an insignificant amount and then
rises further, Fig.3.32a. If the disturbance occurs at
higher loads, one reaches a load level at which the prebuckling load at the moment of disturbance and the highest
postbuckling load are equal, Fig.3.32b. If the disturbance
is produced at a still higher load level, all postbuckling
loads remain below the buckling load, Fig.3.32c.

The buckling load that is exactly equal to the highest
postbuckling load, Fig.3.32b, is the lower limit of the
load-carrying capacity of the disturbed cylinder. It var-
ies little for different cylinders of the same nominal
dimensions.

The relationship between the smallest load of the disturbed
cylinder, which we may call the "disturbed buckling load",
and the load-shortening curve in the postbuckling region
is illustrated in Fig.3.33. Since the cylinders in our test-
set up are loaded in a controlled-deformation condition,
they cannot buckle before the smallest postbuckling shortening is reached on the prebuckling line. In a perfect cylinder the smallest postbuckling shortening is reached with
the characteristic postbuckling curve. As the initial imperfections are much smaller than the postbuckling deformations of axially loaded cylinders, at loads far below the
buckling load, the results of postbuckling calculations
obtained for perfect cylinders can be applied with fairly
good approximation to imperfect cylinders, too. If this
procedure was exact, the lower limit of the experimental
buckling loads would be the prebuckling load which corresponds to the smallest postbuckling shortening of the perfect cylinder, Fig.3.33a. If one accepts the above arguments axiomatically and at the same time wishes to provide
for irregularities of the structure not considered in the
calculations, one may define as lower limit of the experimental buckling loads of disturbed cylinders the postbuckling load associated with the smallest end-shortening, i.e.
the characteristic postbuckling load, Fig. 3.33b.

In order to find out whether it is a reliable lower limit,
we caused a number of cylinders to buckle prematurely by

short time application of local radial point loads and we
found that all buckling loads of the disturbed cylinders
were higher than the characteristic postbuckling load,
Fig. 3.34.

In a perfect cylinder the snap-back from the postbuckling
zone to the prebuckling line happens at the point of lowest
end-shortening perpendicularly upward, Fig. 3.33a. Hence,
the load at which the last buckle snaps out and the lower
limit of the scatter region of the buckling loads are the
same. We measured in a number of tests the smallest buckling load of the disturbed cylinder as well as the load at
which the last buckle snaps out and found that the two
loads approximately coincide.

Loading by controlled shortening with parallelly guided
plates is achieved mainly in a laboratory. For practical
applications, tests under controlled load with the end
plates of the cylinders free to tilt, are more important.
Therefore, we performed buckling tests using a controlled
load device, where the cylinder was loaded in steps of 50 %
weight. The external radial disturbance was produced by a
device that permits measurement of indentation depth and
indentation force. The falling down of the upper bulkhead
of the cylinder during buckling was broken by props so that
the cylinder could not collapse.

Cylinders subjected to controlled shortening will not buckle
before the smallest postbuckling shortening is reached; whereas cylinders subjected to controlled load can snap-through as
soon as the level of the lowest postbuckling load is exceeded,
Fig. 3.33a.

Fig. 3.35 shows buckling loads of a disturbed cylinder under
controlled load as function of the indentation depth which

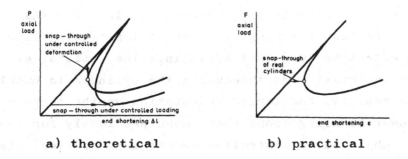

a) theoretical   b) practical

Fig.3.33 Lower limits of experimental buckling loads .

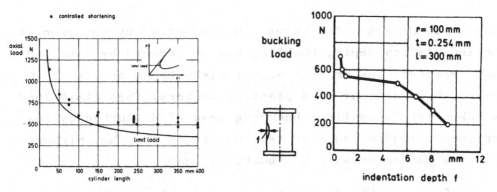

Fig.3.34 Disturbed buck-
ling loads in a deforma-
tion controlled test device .

Fig.3.35 Buckling loads as
function of indentation depth .

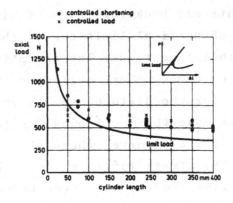

Fig.3.36 Disturbed buckling loads and
characteristic postbuckling
load .

was necessary to stimulate premature buckling. For the evaluation of the load carrying capacity of imperfect cylinders it is important to know that a distinct load exists, at which the indentation depth causing the cylinder to buckle increases rapidly. For larger indentation depths there are indeed lower buckling loads than found previously for the cylinder subjected to controlled shortening. However, disturbances of this size need not be considered as unavoidable, but may be introduced, if necessary, into the stability proof as a prescribed external load.

Only at loads above the threshold, indicated by the jump of the indentation depth, buckling can be induced by small initial imperfections or by a slight disturbance occurring in the rough conditions of a plant or factory. This limit load is taken as the smallest buckling load of disturbed cylinders under controlled load testing; it is somewhat higher than the characteristic postbuckling load.

Fig.3.36 shows all the smallest buckling loads obtained from tests with disturbed cylinders. The scatter regions of the buckling loads of cylinders under controlled shortening and controlled load practically coincide. With the exception of three outlying points all buckling loads of disturbed cylinders lie above the theoretical limit curve, that is above the characteristic postbuckling loads. The exceptional points are buckling loads for extremely short cylinders that failed due to local buckling under the disturbance. It can be certainly assumed that in this case the local buckling loads lie also above the theoretical limit curve, whereas on large parts of the circumference, the stress level is lower. Thus, the characteristic postbuckling load is the lower bound on the scatter region of local experimental buckling loads of disturbed cylinders.

It is a matter of some pride that we succeeded in finding
a postbuckling load that can be regarded as the lower bound
on the scatter region of the smallest buckling loads of a
disturbed cylinder. This lower bound is naturally also the
lower bound of the experimental buckling loads of undisturbed cylinders. But for practical application, i.e. as design load for undisturbed cylinders, it is too conservative.

Hence, the problem of finding a lower bound on the scatter
region of the experimental buckling loads of undisturbed
cylinders still remains.

But from the discussion of the buckling behavior of disturbed cylinders we may have a proper approach to the latter problem. We know

- that the sensitivity to initial imperfections indeed
  depends on the postbuckling behavior and

- that the index value for the lowest snap-through load
  of an axially compressed cylinder is given by the
  smallest postbuckling shortening.

These are positive conclusions. A negative conclusion is

- that the lower bound on the scatter region of experimental buckling loads cannot be found in the stable
  postbuckling region, as the stable postbuckling loads
  are dependent on the cylinder length, while the theoretical and experimental buckling loads are, on the
  contrary, independent of length.

With these three statements we have only one way open to
seek the lower bound on the scatter region of axial buckling loads: we must calculate the smallest postbuckling
shortenings attained in the unstable postbuckling region.

3.2.5 <u>Experimental investigations of unstable equilibrium</u>
<u>states</u>

High-speed motion pictures demonstrated that the stable
postbuckling pattern, visible in buckling tests after the
snap-through, has nothing to do with the buckling behavior.
In the initial postbuckling region the unstable v.Kármán
diamond pattern appears. Only on the basis of this unstable
diamond pattern, the postbuckling load that is a lower bound
on the scatter region of experimental buckling loads, can be
defined. If we may bring in the comparison with the story of
the gold diggers again, we are now seeing the treasure deep
in a ravine, and our problem is the hauling out of it.

To this end, we shall survey the terrain, that is, we should
calmly examine the unstable postbuckling history. Fig.3.37
shows a sampling from high-speed photographs taken of a Mylar
cylinder with the dimensions $r = 100$ mm , $t = 0.254$ mm and
$L = 330$ mm . The numbers in the lower right hand corner of
each frame refer to the time history of the buckling process
from the onset of bucklin in ms . The entire buckling history, from the appearance of the first buckle to the stable
two-tier postbuckling pattern, takes place in about 10 ms .

The first buckles appear simultaneously at two different
spots: in the middle of the cylinder and towards the lower
edge. They are relatively small and have equal extensions in
axial and circumferential directions. Buckles of that kind
will be called square in the following. With some benevolent
intention one may state that the size of the first buckles
corresponds to a circumferential wave number $n = 18$; this
is the identical wave number as resulting from the classical
buckling formula, when square buckling form is assumed.

123

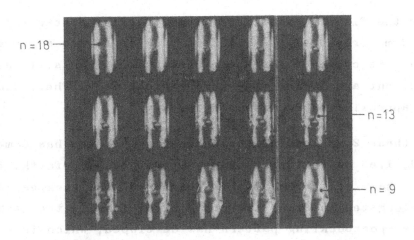

Fig.3.37 High-speed motion pictures of the buckling process
of an axially loaded isotropic cylinder with
dimension r = 100 mm, t = 0.254 mm, 1 = 330 mm .
The pattern of vertical dark zones visible at
time 0 is due to the arrangement of lights.

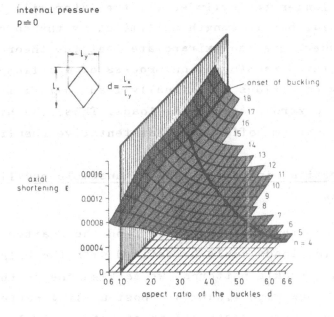

Fig.3.38 Minimum values of axial shortening calculated
for the cylinder, shown in Fig.3.37 .

Around the first buckles, additional buckles appear till
after 2 ms from onset, a field of buckles covers a large
part of the cylinder surface. These buckles are still square
shaped, but are larger than the original ones. Their dimensions approximately correspond to n = 13 .

After these 2 ms the intrinsic buckling process has come to
an end, i.e. no further buckles supervene. Henceforth, the
square shape of the buckles is abandoned. The buckles grow
more elongated in axial direction until finally the stable
two-tier postbuckling pattern has developed, which is well-
known from numerous experimental postbuckling investigations
on axially loaded thin walled cylinders.

The second part of the buckling process, in which the buckles
are only transformed, illustrates the reason why the circumferential wave numbers decrease with increasing cylinder
length. The longer the cylinder is, the later will the
growth of axial buckle length be limited by the endplates
of the cylinder. One can extrapolate that the theoretical
approach of this transformation process of a cylinder without boundary conditions will finally lead to infinitely long
buckles and to zero postbuckling loads. Thus, the high-speed
photographs confirm Hoff's somewhat tentative results.

3.2.6   Theoretical investigations of unstable equilibrium states

As we were sure that the lower bound on the scatter region
of the buckling loads could be defined with the help of the
diamond-shaped buckle patterns, we tackled the postbuckling
calculations anew with v.Kármán's postbuckling pattern. In
contrast to earlier investigations of infinitely long cylinder, we now were not looking for the smallest postbuckling
load, but for the smallest postbuckling shortening.

We had the nebulous hope that it would be possible to find
with the aid of the smallest end-shortenings a postbuckling
load that could be taken as the lower bound on the scatter
region of experimental buckling loads. With this hope we
computed, for many circumferential wave numbers and aspect
ratios of the buckles, the smallest end-shortenings, that
could be attained for infinitely long cylinders with diamond-shaped postbuckling patterns. Fig.3.38 presents the
results of these computations. It shows the smallest end-
shortenings plotted versus the circumferential wave number n
and the aspect ratio d of the buckles.

On this surface the path, run through by the cylinder, has
been traced. The buckling process begins with square buckles,
the circumferential wave length of which corresponds to
n = 18 . Then the circumferential wave number decreases, the
square shape of the buckles being maintained. In this direction, which is emphasized in the diagram by a plane with vertical hatching, the axial shortening gets smaller until a minimum of the shortening is reached at n = 13 . Here the intrinsic buckling process terminates. From then on, the path
guided by decreasing shortening, follows a new line. The
buckles grow more elongated in axial direction: the circumferential wave number decreases further. In our tests the
buckling process was stopped, since the cylinder edges prevented further elongation of the buckles.

It is not immediately convincing that the cylinder loaded
by controlled-shortening changes its shortening continuously
during the buckling process. But the high speed pictures are
so clear and the analysis is so transparent, that their validity cannot be doubted. There are two arguments with which
one can prove that the logic seemingly confused, in reality
is not violated.

● At the beginning of the snap-through the cylinder
buckles only within a narrow region. It is quite
probable that this narrow buckled region is shortened more than on the average, Fig.3.39. During the
snap-through, the buckled region grows larger and
its shortening gets smaller. When the buckling process has ended, the buckled region covers the whole
length of the cylinder and, hence, its shortening
has decreased to the average shortening again.

By the way, Fig.3.39 makes it plausible that the
buckling behavior of imperfect and disturbed cylinders is equal under controlled shortening and controlled load testing. The figure confirms that it
was reasonable to define, as the lower bound on the
scatter region of the buckling loads of disturbed
cylinders, the load corresponding to the smallest
postbuckling shortening rather than the load corresponding to the same shortening on the prebuckling
line.

● For a cylinder subjected to pure axial load in a
deformation-controlled test device, the area under
the load-shortening curve represents the potential
energy. Fig.3.40a shows as an example the area representing the energy for the point of smallest
postbuckling shortening. This point of the smallest
shortening is likewise the point with the lowest
potential energy. Fig.3.40b presents the lowest potential energies as function of the circumferential
wave number n and the aspect ratio d of the buckles. This surface has in the main the same shape as
the surface of smallest shortening, shown in Fig.3.38.
These minimum values of the potential energy may be

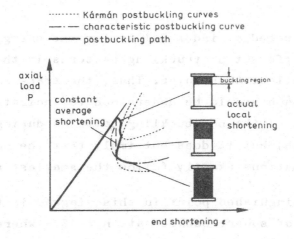

Fig.3.39 Local and average axial
shortening .

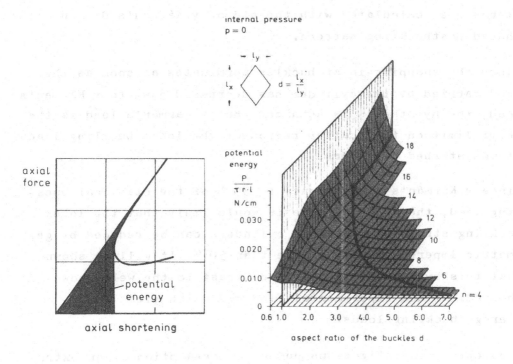

a. Graphic illustration   b. Minimum values

Fig.3.40 Potential energy .

interpreted as index values for the energy level of
the different postbuckling patterns in the region
of smallest shortening. Thus, the trace shown in
Fig.3.40b should be understood to indicate the sequence of the postbuckling patterns during the snap-
through; but it does not imply that the postbuckling
deformations exactly follow the smallest shortening.

The only distinguished point in this diagram is the minimum
on the curve of square buckles at n = 13 , where the intrinsic buckling process terminates. The corresponding axial
load for the isotropic cylinder amounts to 50 % of the classical buckling load. We named it "v.Kármán's load", because
it has been calculated with the aid of v.Kármán's diamond-
shaped postbuckling pattern.

Since the snapping-in of buckles terminates as soon as the
load carried by the cylinder has decreased down to v.Kármán's
load, the hypothesis is advanced that v.Kármán's load is the
lower limit on the scatter region of the local buckling loads
of undisturbed cylinders.

Since v.Kármán's load amounts to 50 % of the classical buckling load, the above hypothesis would imply that the local
buckling stress of isotropic cylinders can be reduced by geometric imperfections not more than 50 %. It will be shown
that this hypothesis is not in contrast to the well-known
theoretical and experimental findings which yield lower
average buckling loads.

Theoretical investigations under the assumption of an extra-
ordinarily unfavorable imperfection pattern yielded a much
greater reduction. It is improbable however, that such an
unfavorable regular imperfection pattern is produced by
chance.

Experimental investigations have shown that very thin-walled cylinders frequently buckle under loads which are below v.Kármán's load. But these experimental buckling loads are not the actual local loads, but are average values. It is obvious that the establishment of a lower limit for the load carrying capacity of an axially loaded cylinder must rely on local stress, since it is well known from numerous buckling and postbuckling tests that cylinders with extremely small buckling loads always buckle locally.

Stress peaks may occur due to three causes: Residual stresses, geometric imperfections and irregular load introduction.

About residual stresses nothing can be said because extensive theoretical or experimental investigations are not known to the authors.

Geometric imperfections and their growth in the prebuckling region have been measured by Arbocz [18] for copper cylinders with dimensions r = 101,6 mm, t = 0.1153 mm and l = 203.1 mm. Fig.3.41a shows the pattern of initial imperfections; it can be seen that the imperfection modes with one axial half wave have the largest amplitudes. Fig.3.41b illustrates the prebuckling deformation growth just before buckling; it reveals that under increasing load the modes with one half wave grow predominantly. It should be remarked that the plots do not cover the whole length of the cylinder, since at the edges no measurements could be taken. The deformations in the prebuckling region lead to an irregular stress distribution rendering the cylinder liable to buckling. The high speed motion pictures showed that the buckling process always begins with the snapping-in of small square buckles (cf. Fig.3.37). Hence, the long wave prebuckling pattern shown in Fig.3.41b does not lead to a limit point.

Irregularities of load introduction are particularily deleterious for thin-walled and short cylinders. Their critical
end-shortening

$$\Delta l = \varepsilon \cdot l = 0.6 \frac{t \cdot l}{r}$$

is small and, hence, they are particularily sensitive to out
of planeness of the edge support.

In Fig.3.36 three outlying points, i.e. three buckling loads
of short cylinders falling below the curve of characteristic
postbuckling loads, have attracted attention. We interpreted
this exception with the assertion that for these points the
v.Kármán load was exceeded locally, whereas on large parts
of the circumference the stress level was lower.

The hypothesis that the local buckling stress of axially
loaded cylinders is reduced by random deviations from the
perfect shape, not more than 50 % of the classical buckling
load should be checked by measuring the stress in the cylinder before buckling.

Preliminary tests have shown that with our Mylar cylinders
the buckling loads could be reduced remarkably by placing
a cigarette paper under the cylinder edge flange. On the
other hand the buckling load of an irregularly buckling cylinder could be increased by 10 %, if some thin foil was put
in at the region where the last buckles had disappeared previously.

3.2.7 Concluding remarks

In the foregoing chapter it was outlined that the lower limit
on the local buckling loads of isotropic circular thin-walled
cylinders under axial compression is v.Kármán's load, which
is defined as the load corresponding to the smallest axial

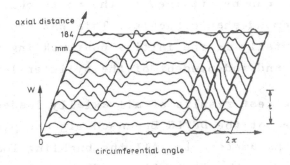

a. Initial imperfections

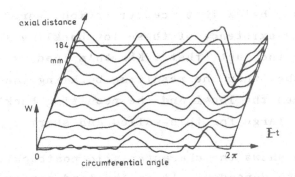

b. Prebuckling deformation growth

3.41 Radial deviations from the perfect shape .

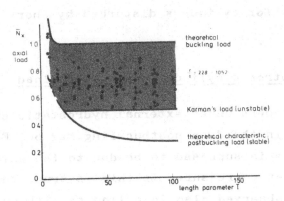

Fig.3.42 Scatter region of the experimental buckling loads.

shortening that can be attained in the postbuckling region
with square diamond-shaped buckles. This pattern has been
found experimentally in the initial postbuckling region;
it is unstable and independent of the cylinder length.

In Fig.3.42 the test results of all axially loaded isotropic
cylinders tested at the DFVLR Braunschweig are plotted versus the length parameter $\bar{l}$. All the buckling loads lie in
the scatter region, the lower limit of which is v.Kármán's
load.

Buckling loads, below that scatter region, were measured
elsewhere. The existence of these low buckling loads may be
explained, by the fact, that v.Kármán's load is the limit on
the local rather than the resulting buckling loads. The difference between the local and the resulting buckling load is
particularily large for thin-walled and short cylinders.

Fig.3.42 also shows the characteristic postbuckling load
which is length dependent. It is the load corresponding to
the smallest shortening attained with a stable postbuckling
pattern. There is no relation between this load and the experimental buckling loads. The characteristic postbuckling
load may be of some importance as the lower limit on the
buckling loads for cylinders disturbed by short time radial
loads.

3.3 Short isotropic cylinder under axial load

Isotropic cylinders under external hydrostatic pressure
undergo twist in the deep postbuckling region, Fig.3.5. This
mode of failure is supposed to be due to the axial component
of the hydrostatic pressure. If this is true, the torsional
mode would be observed also in cylinders with pure axial
load. But in our tests, reported above, no loss of stability

by torsion has occurred. First, we believed the reason of
this to be the rigid support of the cylinders in our test
device, and we imagined that we should find torsional postbuckling patterns, if the restriction of torsion would be
removed by supporting the test cylinders on spheres. But
the first test specimens supported in this manner were not
inclined to twist when subjected to axial compression.

We found torsional modes only, when we turned to investigate
cylinders, which were so short that at buckling they snapped
through into one-tier postbuckling patterns. As an example
we show the test results obtained with a cylinder, the dimensions of which are radius r = 200 mm , wall-thickness
t = 0.254 mm and length ι = 35 mm . The lower endplate
of the cylinder was supported by a ball-bearing which al-
lowed torsion and lateral shift.

In the course of loading the postbuckling pattern appeared
locally. It consisted of one row of buckles without torsion.
Fig.3.43 shows the load shortening curve. The load at which
the first buckles snapped in is designated as the experimental buckling load. It amounts to 80 % of the theoretical
buckling load of the perfect structure.

With increasing shortening the postbuckling pattern was completed while the load increased further, but at a lower rate.
The completed postbuckling pattern, which is still a one-tier
pattern without torsion is shown in Fig.3.44a.

At further shortening the load attained a maximum. There the
cylinder by twisting withdrew from carrying higher loads. The
torsional postbuckling pattern is shown in Fig.3.44b. Because
of its graceful swinging appearance we called it "Kreutzberg
Pattern" after the famous dancer Harald Kreutzberg.

134

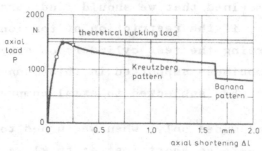

Fig.3.43   Load-shortening curves.

a. Without torsion

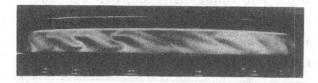

b. Kreutzberg pattern

c. Banana pattern

Fig.3.44   Postbuckling patterns of a very short cylinder
r = 200 mm,   t = 0.254 mm,   l = 30 mm .

When the shortening was increased further the load decreased
slowly until the postbuckling pattern suddenly changed into
the "Banana Pattern" shown in Fig.3.44c. At this event, which
was acoompanied by a load report, the load dropped considerably.

With further shortening the load remained nearly constant. The
snapping-in of the banana pattern had affected the cylinder so
strongly that it would have been meaningless to continue the
shortening furthermore.

It is remarkable that the experimental buckling load is considerably lower than the theoretical buckling load of the perfect cylinder, although the postbuckling behavior is such that
a lowering influence on the buckling load should not be expected; small postbuckling loads exist indeed, but are connected with high values of shortening and thus do not affect
the onset of buckling. The low experimental buckling load is
due to an irregular introduction of the axial force. This ir-
regularity is unavoidable with short cylinders.

3.4 Isotropic cylinders under torsion

The following statements on thin-walled isotropic cylinders
under torsion are based on a paper by Yamaki [19] presented
at the IUTAM Symposium at Harvard University in Cambridge in
1974. The paper reports on carefully conducted experiments
and moreover contains a survey of earlier research work on
the subject matter.

The test specimens were Mylar cylinders with radius $r = 108$ mm,
wall-thickness $t = 0.254$ mm, Young's modulus $E = 5670$ N/mm^2
and Poisson's ratio $\nu = 0.3$. The cylinder lengths were var-
ied; the specimens had lengths equal to 23, 36, 51, 71, 114
and 161 mm respectively. In the figures the Batdorf parameter
is used, defined through

136

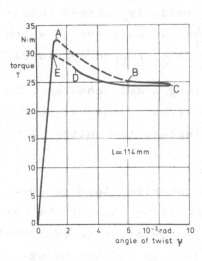

Fig.3.45 Torque-angle of twist-curve r = 108 mm , t = 0.254 mm, l = 114 mm .

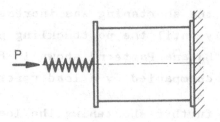

Fig.3.46 Schematic presentation of a soft test device .

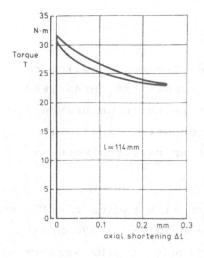

Fig.3.47 Torque-end shortening curve .

Z = 100

Z = 1000

Fig.3.48 Postbuckling patterns .

$$Z = \sqrt{(1-\nu^2)} \; \frac{l^2}{r \cdot h} \; .$$

The test specimens correspond to Z values of 20, 50, 100,
250, 500 and 1000 respectively.

Load-deflection curves extending into the advanced postbuckling region are presented only for the cylinder of length
l = 114 mm . Fig.3.45 shows the relation between the torque
T and the angle of twist ψ . The point A means the buckling load, the curves A-B and D-E correspond to the unstable snap-through in the loading and unloading period respectively, and the curve B-C-D represents stable equilibrium states in the postbuckling region.

The testing device was nominally deformation controlled, but
actually it was so weak that the deformation by no means remained constant during the snap-through. The behavior of a
deformation controlled weak test apparatus will be illustrated with aid of Fig.3.46. In the prebuckling region the
arrow moves slowly forward under controlled deformation,
till the cylinder buckles. At the instant of buckling the
arrow remains stationary. The load carried by the cylinder
drops. Hence, the spring is partially unloaded and consequently expands causing further shortening of the cylinder,
although during this period the arrow remains fixed in its
position.

We turn to Fig.3.45 once more and remark that during the
snap-through A-B the torsional moment drops and the twist
becomes large. In the postbuckling region B-C the equilibrium torque decreases monotonically with increasing twist.
The hysteresis loop in the range B-C-D indicates a slight

friction to the movement of the cylinder - according to Yamaki.

The relation between the torque and the axial shortening is given in Fig.3.47. It can be seen that in the postbuckling region the length of the cylinder decreases monotonically with the development of the buckles.

Fig.3.48 shows postbuckling patterns of cylinders with length parameters Z = 100 and Z = 1000. One sees that the cylinder buckles with one half wave axially and a number of full waves in circumferential direction. The waves are oriented obliquely with respect to the torque axis.

Postbuckling patterns with one half wave in axial direction have been observed also on cylinders under external pressure. This similarity of the postbuckling patterns involves a number of common features in the buckling and postbuckling behavior of cylinders under torsion and under external pressure, respectively:

- the longer the cylinder, the smaller the buckling and postbuckling loads,

- the longer the cylinder, the smaller the circumferential wave number,

- the circumferential wave number does not change spontaneously in the postbuckling region,

- postbuckling equilibrium states with circumferential wave numbers smaller than the initial one, can be produced artificially in the advanced postbuckling region.

To clarify the influence of shell geometry on the buckling and postbuckling behavior of cylinders under torsion the

torque-twist curves for cylinders with various length parameters Z are plotted together in Fig.3.49, upper frame.
In this figure T_{cr} is the critical torque theoretically
predicted, while ψ_{cr} is the corresponding angle of twist.
It is interesting to note that the experimental buckling
loads amount to approximately 90 % of the theoretical ones.
The postbuckling curves rise in the case of short cylinders,
but drop in the case of long ones. - Cylinders under external
pressure and under axial load principally behave in the same
way; you have seen that in the preceeding chapters and you
will see it more comprehensively in the next chapter.

Theoretical postbuckling investigations on cylinders under
torsion have been performed for the initial postbuckling
region only. The lower frame of Fig.3.49 represents theoretical postbuckling curves published by Budiansky [20] in
1969. At the first glance there seems to be good agreement
between experimental and theoretical results, since in both
diagrams the postbuckling curves in the initial postbuckling
region diverge in a fan-shaped way. But on closer inspection
essential differences will be noticed. The theoretical curves
exhibit a distinct minimum of the slope at Z = 30 , whereas
the experimental curves indicate a monotonic decrease of the
slope with increasing parameter Z . It can not be excluded
that the theoretical results might be correct for perfect
cylinders, but actual cylinders behave in a quite different
manner, probably due to the influence of initial imperfections. Since the strong drop of slope theoretically predicted
for cylinders in the region of Z = 30 is not noticable in
the experiments, it can not have a remarkable influence on
the imperfection sensitivity. We shall discuss a similar
situation in section 4.2.4 for spherical shells.

In Fig.3.49 the torsional moment is related to the theoretical

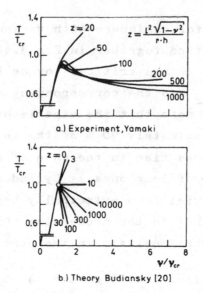

a.) Experiment, Yamaki

b.) Theory Budiansky [20]

Fig. 3.49 Torque-angle of twist-curves in the
initial postbuckling region .

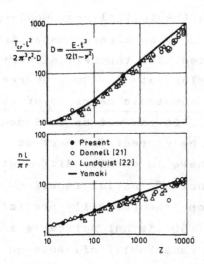

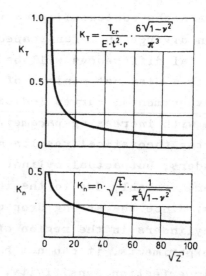

Fig. 3.50 Torsional buckling
loads in logarithmic scale .

Fig. 3.51 Torsional buckling
loads in natural scale .

critical torque. If one wants to know the actual values of the torsional moments one needs the critical torques. These are presented in Fig.3.50, upper frame, and the corresponding circumferential wave numbers in Fig.3.50, lower frame. It can be seen that the experimental buckling loads and wave numbers agree fairly well with the theoretical predictions.

The representation as utilized in Fig.3.50 has the advantage that a large region can be exhibited graphically, but has the disadvantage that one can not directly recognize the influence of any particular parameter. In order to clearly bring out the dependence of the critical load and the corresponding circumferential wave number on the cylinder length we give a transformed representation of Fig.3.50 in Fig.3.51. In this figure one sees that the loads and the wave numbers decrease with increasing cylinder length as in the case of external pressure.

3.5 Orthotropic cylinders

3.5.1 Introduction

The theoretical postbuckling investigations on isotropic cylinders as outlined in the chapters 3.1 and 3.2 were rather extensive. The question may arise, whether these efforts were worthwile. This seems at first sight unlikely, considering the fact that the results of a lot of buckling and postbuckling tests on isotropic cylinders are available and that postbuckling analysis can hardly be considered to improve either the economy or safety of such structures. Despite this, the theoretical postbuckling investigations on isotropic cylinders are important, not for their own sake, but as preparation for the optimization of stiffened cylinders.

The load carrying capacity of thin-walled cylinders under

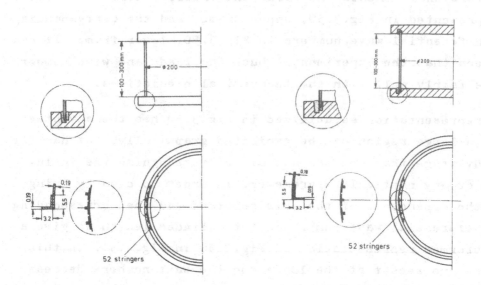

a. Stringers outside   b. Stringers inside

Fig.3.52   Test specimens .

a   b

a. Outside-stiffened   n = 8   b. Inside-stiffened   n = 8

Fig.3.53   Test-cylinders buckled under
           axial load   l = 130 mm .

compression can be markedly improved by adding stiffeners,
which contribute little to the weight. For the optimization
of stiffened cylinders, it is not sufficient merely to compare bifurcation loads of perfect cylinders, but the sensitivity toward initial imperfections has to be considered too.
For taking into account the imperfection sensitivity, we will
now fall back on the postbuckling investigations of isotropic
cylinders. There, for the lower limit of the scatter region
of experimental buckling loads we had defined, postbuckling
loads which could be calculated for the perfect cylinder.
This lower limit consists for sufficiently long cylinders of
v.Kármán's load and for short cylinders of the characteristic
postbuckling load. The definitions of these limit loads shall
be assigned to orthotropic cylinders as well. In the design
of orthotropic cylinders they will be used as index values
for optimization.

Thus the procedure of optimizing orthotropic cylinders is
clearly formulated and could be accomplished. But since the
capacity of our institute is not sufficient to tackle all
fascinating problems at once, we have done only the first
steps in this direction up to now. We can report to you
about our results of experimental and theoretical investigations on longitudinally stiffened cylinders only [23,24].

The test cylinders were made out of Mylar foil. They consisted of an isotropic skin and 52 Jᒋ shaped stiffeners glued
on. Fig.3.52 shows the dimensions of the test specimens and
the arrangement of the stiffeners. The stiffeners were lo-
cated so close to each other that for the calculation their
rigidity could be taken as equally distributed over the surface of the shell. Four cylinders were produced, two with internal and two with external stringers. After each test the
cylinders were cut to a smaller length and then retested

because we wanted to investigate the influence of the cylinder length upon the buckling and postbuckling behavior.
The graduation of length was l = 300, 260, 295, 130 and
100 mm respectively.

All cylinders after buckling snapped into a one-tier pattern,
under external pressure as well as under axial compression.
Fig.3.53 shows some examples.

3.5.2 Cylinder under external hydrostatic pressure

Fig.3.54 presents the theoretical and experimental buckling
pressures of the internally stiffened cylinder as function
of the cylinder length. In addition, the circumferential
wave numbers for the theoretical buckling patterns and of
the experimental postbuckling patterns are indicated in the
figure. With the good agreement, between the theoretical and
experimental buckling loads, one can come to the conclusion
that the cylinder is not sensitive to initial imperfections
and that the postbuckling loads are probably not much smaller
than the buckling loads.

Fig.3.55 shows the load-shortening curve for an internally
stiffened cylinder of length 300 mm. It can be seen that the
characteristic postbuckling load is of the same magnitude as
the theoretical buckling load. A reduction of the buckling
pressure by initial imperfections is therefore hardly possible. The fact that the experimental buckling load is even
a bit larger than the theoretical one, may have its root in
the scatter of the test specimens relative to wall-thickness
and Young's modulus.

Fig.3.56 exhibits, as function of the circumferential angle,
the radial deviations from the circular shape, measured in
the middle of the cylinder in unloaded condition, shortly

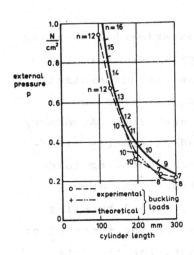

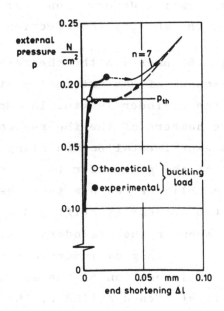

Fig.3.54 Buckling loads of internally stiffened cylinders under external pressure .

Fig.3.55 Load-shortening curve of an inside stiffened long cylinder l = 300 mm .

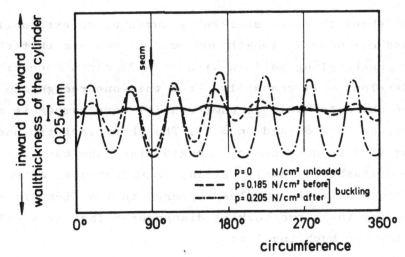

Fig.3.56 Radial deviations of the ideal shape for an inside stiffened cylinder under external pressure l = 300 mm .

prior to buckling and after buckling. One can see that the
postbuckling deformations are rather regular and much bigger
than the initial imperfections.

Fig.3.57 presents the theoretical and experimental buckling
pressures of the externally stiffened cylinder as function
of the cylinder length. In addition, the circumferential
wave numbers of the theoretical buckling patterns and of
the experimental postbuckling patterns are given. A comparison of the graphs for the internally and externally stiffened cylinder reveals that the theoretical buckling loads
of the externally stiffened ones are higher, the more so
the shorter the cylinders are. The experimental buckling
loads for long cylinders are independent of the internal or
external location of the stringers. Hence, for the externally stiffened cylinder, the reduction of the buckling load
by initial imperfections is bigger. This gives us cause to
suppose that for externally stiffened cylinders the characteristic postbuckling load is probably below the theoretical
buckling load.

Fig.3.58 shows the load shortening curve of an externally
stiffened cylinder of length 300 mm. One can see that the
regular postbuckling pattern with $n = 7$ circumferential
waves develops only gradually. From this one recognizes
that the characteristic circumferential wave number is at
the limit of $n = 8$ and $n = 7$. The characteristic postbuckling load is not uniquely marked since the buckling process is disturbed by small initial imperfections. After the
snap-through only two buckles appeared in the first instance;
on unloading these two buckles disappeared last at a relatively low postbuckling load.

Such a buckling and postbuckling behavior is frequently observed in cylinders subjected to external pressure. At this

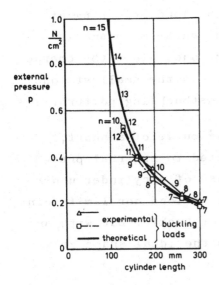

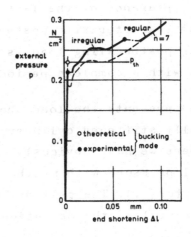

Fig.3.57 Buckling loads of externally stiffened cylinders under external pressure.

Fig.3.58 Load-shortening curve of an outside stiffened long cylinder l = 300 mm.

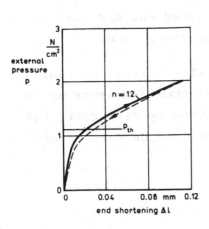

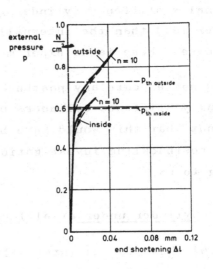

Fig.3.59 Load-shortening curve of an outside stiffened short cylinder l = 100 mm.

Fig.3.60 Load-shortening curves of an outside and an inside stiffened cylinder l = 130 mm.

point it should be emphasized that the load occurring at
the disappearance of the last buckle can be much smaller
than the characteristic postbuckling load, since the characteristic postbuckling load is defined as the smallest load
reached with a complete periodical postbuckling pattern.

Fig.3.59 presents the load-shortening curve of a short,
externally stiffened cylinder subjected to external pressure. Here we have obviously the case of a cylinder under
external pressure with neither buckling load nor limit point,
cf. chapter 3.1. The cylinder does not buckle. But with an
increase of load the deviations from the ideal shape grow
larger.

Fig.3.60, the last figure of the series on stringer stiffened cylinders under external pressure, demonstrate the difference in the load shortening curves of an internally and ·
externally stiffened short cylinder. One recognizes that the
externally stiffened cylinder gets finite deflections at a
greater load than the internally stiffened one and that in
both cases there is no experimental buckling load.

We did not execute any postbuckling calculations for longitudinally stiffened cylinders under external pressure up to
now; not that this would have been extremely difficult, but
other postbuckling investigations seemed to be more interesting to us.

3.5.3 Cylinder under axial load

Fig.3.61 shows for an internally stiffened cylinder under
axial load, as functions of the cylinder length: the theoretical buckling loads, the experimental buckling loads, the
minimum postbuckling loads and the loads at which the last
buckle disappeared. The load at which the last buckle jumps

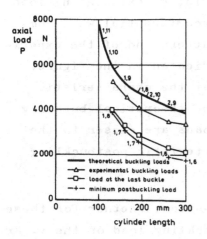

Fig.3.61 **Buckling and post-buckling loads of inside stiffened axially loaded cylinders .**

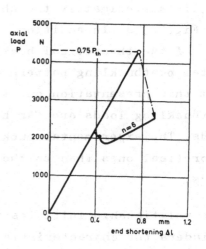

Fig.3.62 **Load-shortening curve of an inside stiffened long cylinder l = 300 mm .**

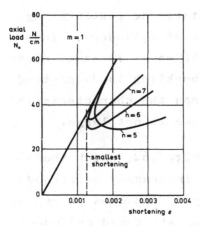

Fig.3.63 **Theoretical post-buckling curves of an inside stiffened cylinder l = 300 mm .**

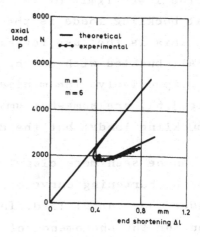

Fig.3.64 **Load-shortening curves of an inside stiff-ened cylinder l = 300 mm .**

out, is approximately the characteristic postbuckling load,
cf. Fig.3.33a. In addition, the circumferential wave numbers of the theoretical buckling pattern and of the experimental postbuckling pattern are indicated in the figure.
From this presentation one sees that the characteristic
postbuckling loads are far below the theoretical buckling
loads. The experimental buckling loads are closer to the
theoretical ones than to the characteristic postbuckling
loads.

This relationship initiates the question, whether for these
cylinders the characteristic postbuckling load or the v.Kár-
mán load is the lower limit on the scatter region of the experimental buckling loads. The answer can easily be found by
a crictical inspection of the theoretical buckling patterns.
If the cylinder is so short or so strongly stringer stiffened that there cannot develop a v.Kármán diamond shaped
postbuckling pattern, uninfluenced by the cylinder edges,
then the lower limit of the scatter region of the experimental buckling loads is the characteristic postbuckling
load. This is the case for all our test cylinders, since
they all buckled with one or two half waves in axial direction respectively. The minimum postbuckling loads plotted
in Fig.3.61, are somewhat smaller than the characteristic
postbuckling loads; but the difference is not large.

That can be seen more clearly from Fig.3.62, which shows
the load shortening curve of a long internally stiffened
cylinder under axial load. This diagram provokes an explanation for the phenomenon of stringer stiffened cylinders
having high experimental buckling loads despite their low
postbuckling loads. In response to this challenge we state
that for stiffened cylinders the experimental buckling loads

are less reduced than for the isotropic cylinder treated in
section 3.2 for three reasons:

- First, provided the depth of the initial imperfections
  is the same for the stiffened cylinders as for the
  more thin-walled isotropic ones, the ratio

  <u>depth of initial imperfections</u>
  wall-thickness

  is smaller for the stiffened cylinder and therefore
  the initial imperfections are less deleterious.

- Second, for the stiffened cylinders the extension of
  the buckles is larger than for the more thin-walled
  isotropic ones. Provided the depth of the initial imperfections is the same in both cases, the ratio

  <u>depth of initial imperfections</u>
  wave length of the buckles

  that means the curvature due to the initial imperfections, is smaller for the stiffened cylinders and,
  hence, the imperfections are less offensive.

- Third, for the stiffened cylinders the buckling stress
  and, hence, the end-shortening are larger than for the
  more thin-walled isotropic ones. Provided the uneveness
  of the support is the same in both cases, the irregu-
  larity of the load introduction is smaller for the
  stiffened cylinder.

Concluding, one can say that stiffened cylinders are less
thin-walled and, hence, less imperfection sensitive than isotropic ones.

In practical optimization problems the radius of the cylinder
and the load to be carried, generally are prescribed. With

152

this the wall-thickness ratio can be established to a degree
where possible deviations and consequently possible differences of the ratio

$$\frac{\text{experimental buckling load}}{\text{characteristic postbuckling load}}$$

do not enter essentially the optimization. Hence, it will be
sufficient to optimize the characteristic postbuckling loads.
Similar arguments apply for optimization based on v.Kármán's
load.

Fig.3.63 presents calculated load-shortening curves of the
long internally stiffened cylinder. It can clearly be seen
that the characteristic postbuckling pattern has n = 6
waves on the circumference. That agrees with the experimental results, shown in Figs.3.61 and 3.62.

In Fig.3.64 three characteristic postbuckling curves of nominally equal cylinders are shown; one of the curves has been
calculated for the perfect cylinder, the two others are test
results. The agreement is good.

Fig.3.65 exhibits for the externally stiffened cylinder, as
function of the cylinder length: the theoretical buckling
loads, the experimental buckling loads, the minimum postbuckling loads and the loads at which the last buckle disappeared. The theoretical buckling loads are essentially
higher than those shown in Fig.3.61 for the internally stiffened cylinder. The experimental buckling loads of the internally and externally stiffened cylinder differ to a less
degree. This leads to the conclusion that the externally
stiffened cylinder is more sensitive to initial imperfections than the internally stiffened one.

Fig.3.66 shows the experimental characteristic load-shortening

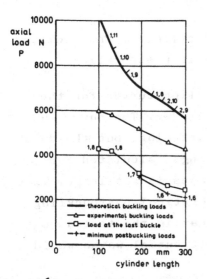

Fig.3.65 Buckling and postbuckling loads of outside
stiffened axiall loaded
cylinders .

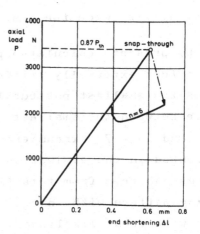

Fig.3.66 Load-shortening
curve of an outside stiffened long cylinder l = 300 mm.

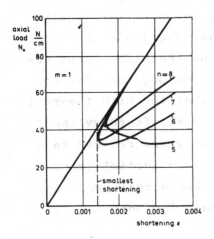

Fig.3.67 Theoretical postbuckling curves of the outside stiffened cylinder
l = 300 mm .

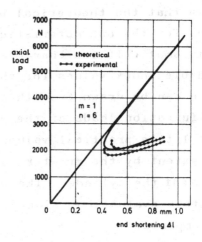

Fig.3.68 Load-shortening
curves of the outside stiffened cylinder l = 300 mm .

curve of a long externally stiffened cylinder. The buckling
load is almost twice the characteristic postbuckling load.

Fig.3.67 presents calculated postbuckling curves for the
long perfect externally stiffened cylinder. The curve n = 7
reaches the smallest postbuckling shortening. But the difference between the smallest shortenings calculated for
n = 6 and n = 7 circumferential waves is not large. Those
experienced in the evaluation of theoretical postbuckling
diagrams can read from this figure that in the test a postbuckling pattern with n = 6 circumferential waves would
appear with great likelihood.

Fig.3.68 shows the comparison of the theoretical and experimental characteristic postbuckling curves. The deviations
are within the scatter of the tests.

In Fig.3.69a the theoretical postbuckling curves of the internally and externally stiffened cylinder are compared. It can
be seen that the theoretical buckling loads differ widely,
whereas for the characteristic postbuckling loads the difference is small. If the characteristic postbuckling loads are
dependable index values for the optimization, then the conclusion can be drawn from this graph that the high theoretical bifurcation loads of the externally stiffened cylinder
are of little practical value, since they are reduced to a
great extent by the greater sensitivity to initial imperfections of the cylinder. The wall-thickness ratio does not
enter this evaluation, since it is equal for both the cylinders compared.

Fig.3.69b shows the experimental buckling loads, together
with the theoretical postbuckling curves known already from
Fig.3.69a. The experimental buckling loads of the internally
and externally stiffened cylinder differ from each other as

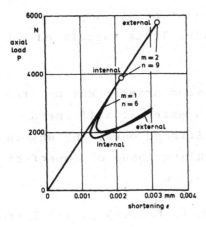

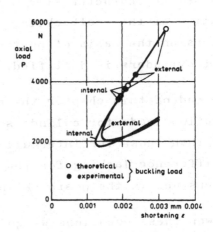

a. Theoretical   b. Experimental and theoretical

Fig.3.69 Comparison of the buckling and postbuckling behavior
of internally and externally stiffened axially loaded cylinder

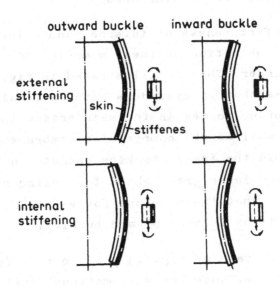

Fig.3.70 Demonstration of the influence of inside or
outside stiffening on the buckling deformation .

little as the characteristic postbuckling loads. Hence, the
evaluation of the postbuckling behavior of the imperfect
cylinders on the basis of postbuckling loads calculated for
perfect cylinders is justified.

At the end of this chapter the question arises: Why are the
bifurcation loads for cylinders with external stiffeners
higher than those for internally stiffened ones, and why is
this difference lacking for the buckling loads of imperfect
cylinders and in the postbuckling region?

To answer these questions we go back to Fig.1.3 in the Introduction. From the left hand lower graph you can see that compressive forces are acting upon the inside buckles and tensile forces upon the outside buckles. The compatibility condition requires that for the membrane forces in axial direction the same applies, i.e. compression for the inside buckles and tension for the outside ones.

The different effectiveness of internal and external stiffening is due to the action of these membrane forces in the
skin of the cylinder. This is illustrated in Fig.3.70. The
upper graph shows that in cylinders with external stiffeners
the tensile membrane forces in the skin acting in the outward buckles as well as the compressive membrane forces in
the skin acting in the inward buckles operate in the restoring sense. The lower graph shows for cylinders with internal stiffeners that the membrane forces act deflecting in
the inward as well as in the outward buckles.

On the right hand frame of Fig.1.3 you see that for the postbuckling region, i.e. once the deformations have rached finite values there are no more alternating membrane forces
in circumferential as well as in axial direction. Hence,
the restoring effect of the membrane forces is gone. Since

the different buckling behavior of internally and externally stiffened cylinders is due to the action of the membrane forces, this difference disappears with the vanishing
of the alternating membrane forces in the postbuckling region.

Initial imperfections act principally in the same manner as
postbuckling deformations do, for they are also finite radial deflections. Hence, they also affect the alternating
membrane forces and consequently cancel the difference in
the buckling loads of internally and externally stiffened
cylinders.

The foregoing discussion affords the opportunity to empha-
size the following principal remark: the sensitivity of
thin-walled shells toward initial imperfections is based
upon the fact that perfect shells are stiffened at the bifurcation point by membrane forces and that for imperfect
shells this stiffening is decreased by small radial deflections. For columns or plates there are no restoring membrane
forces at the bifurcation point, and consequently there is
no imperfection sensitivity due to the failure of these restoring forces.

3.6 Discretely stiffened cylinders

3.6.1 Preliminary remarks

The simplest method of theoretically dealing with the buckling and postbuckling behavior of stiffened shells consists
of uniformly distributing the stiffness of the stringers
and rings over the shell surface. This simplified approach
results in somewhat higher values of buckling loads than
would be obtained by investigation of the actual system.
The deviations become greater, the larger the distance
between the stiffeners.

On the cylinders, whose buckling and postbuckling behavior
will be dealt with in this chapter, the stiffeners are so
far apart that one has to consider the behavior of a discretely stiffened cylinder. The buckling process starts
with local buckling of the panels. The stiffeners get in-
volved in the pattern of large deformations only in the
advanced postbuckling region.

The theoretical investigations of the load carrying capacity
of discretely stiffened cylinders with local buckling of
panels preceding the large deformation of stiffeners are
still in a stage of infancy. Hence, we shall confine our-
selves in the following to discuss only experimental results [25].

The material of our test cylinders was Mylar, a plastic foil
capable of relatively large elastic deformations. The stiffeners consisted of angle sections bonded to the skin, Fig.
3.71. The height of the stiffeners was 1/15 of the cylinder
radius; their weight was less than 20 % of that of the unstiffened cylinder. The edges of the cylinder had been cast
into rigid end plates.

3.6.2 Axially loaded cylinders

For axially loaded cylinders discrete stiffeners may be required for the introduction of concentrated loads.

Fig.3.72 shows load shortening curves which have been re-
produced from records obtained in the experiments. Presented
are the load-shortening curves of four different cylinders:
an unstiffened cylinder, a ringstiffened cylinder, a stringer
stiffened cylinder and a ring- and stringer-stiffened cylinder.

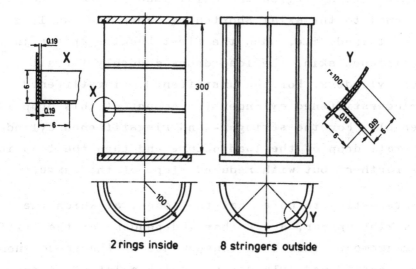

a. for axial load

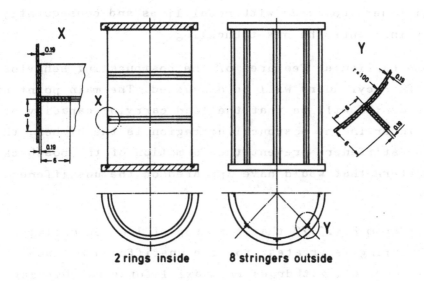

b. for external pressure

Fig.3.71 Discretely stiffened test specimens .

In the prebuckling region the axial load increases nearly proportional to the axial shortening. When the buckling load is attained, i.e. when the first buckles appear in the unstiffened skin, the load drops suddenly for all types of cylinders. For the unstiffened, ringstiffened and stringerstiffened cylinder the amount of this drop is considerable. For the stringer- and ringstiffened cylinder only a small drop of the load occurs and then the load increases further, but with reduced slope of the curve.

It is interesting to note that the loads, at which the first buckles appear, are rather independent of the stiffener arrangement. This is a consequence of the phenomenon, that the unstiffened cylinder has quite a lot of eigen-patterns with loads practically equal to the buckling load. Among this number of patterns there were some for which the stiffeners coincide with nodal lines and consequently did not influence the onset buckling.

Now some particular features of the postbuckling behavior of the four cylinders will be discussed. The main point of the discussion will be that the load carrying capacity of the cylinder in the postbuckling region is the higher, the more the stiffeners prevent the formation of the postbuckling pattern that would have appeared on the unstiffened cylinder.

One sees from Fig.3.72 that the drop of load occurring after buckling is greatest for the unstiffened cylinder; the cylinder wall withdraws from axial force by snapping through into a postbuckling pattern with large radial deformations. This pattern was of the two-tier type observed most frequently on axially loaded cylinders, Fig.3.73b.

In Fig.3.74 you see the ringstiffened cylinder under axial

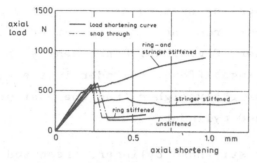

Fig.3.72 Load-shortening curves of an unstiffened cylinder and of three discretely stiffened cylinders under axial compression .

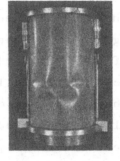

a. one tier   b. two tier
   pattern   pattern

Fig.3.73 Postbuckling patterns of an isotropic cylinder under axial load .

Fig.3.74 Postbuckling pattern of a ring stiffened cylinder under axial load .

 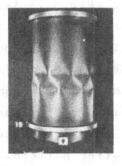

Fig.3.75 Postbuckling patterns of a stringer stiffened cylinder under axial load .

Fig.3.76 Postbuckling patterns of a stringer and ring stiffened cylinder under axial load .

load. The postbuckling pattern of this cylinder is only
slightly different from the two-tier pattern of the unstiffened cylinder, Fig.3.73b. Consequently the axial load
carried by the ringstiffened cylinder in the postbuckling
region is only slightly higher than the postbuckling load
of the unstiffened cylinder.

For the stringer stiffened cylinder, presented in Fig.3.75,
the formation of the two-tier postbuckling pattern is prevented by the stringers. Consequently, a one-tier pattern
appears immediately after buckling, Fig.3.75 left hand
frame, and is maintained up to large values of axial shortening, Fig.3.75 right hand frame. The one-tier postbuckling
pattern of the unstiffened cylinder, shown in Fig.3.73a,
was produced artificially at a load that was somewhat smaller
than the buckling load of the undisturbed cylinder. The postbuckling loads connected with the one-tier pattern are higher
than those connected with the two-tier pattern. Hence, the
postbuckling loads of the stringer stiffened cylinder are
higher than those of the ringstiffened and of the unstiffened cylinder.

Fig.3.76 shows the stringer- and ringstiffened cylinder in
the postbuckling region. In this case the stiffeners disturb
the postbuckling pattern of the unstiffened cylinder in various ways. Immediately after the snap-through, Fig.3.76 left
hand frame, a one-tier pattern is seen. The extension of the
buckles in axial direction is restrained by the rings. In the
panels of the center bay one and two buckles occur alternately. This indicates that the critical circumferential
wave number of a cylinder, the length of which is equal to
the ring spacing, does not agree with the number of the
stringers. In the advanced postbuckling region a three-tier
pattern develops out of the original one-tier pattern,

Fig.3.76 right hand frame. The buckles grow larger and deeper so that two buckles in one panel are no more possible, and the stringers become included in the postbuckling pattern. These various disturbances of the postbuckling pattern that would appear on the unstiffened cylinder are so effective that the postbuckling loads of the stringer- and ringstiffened cylinders are much higher than those of the cylinders stiffened in one direction only.

Summarizing the test results obtained with discretely stiffened cylinders under axial compression it can be stated that

- panel buckling is practically unaffected by the stiffeners. The reason is that for an unstiffened cylinder there exist several eigenpatterns associated with the same buckling load, and from this large choice obviously one pattern, i.e. one set of nodal lines will coincide with the stiffener pattern and, hence, the buckling loads are not changed;

- stiffeners in both directions effectively increase the load carrying capacity of the cylinders as they strongly disturb the postbuckling pattern that would appear on the unstiffened cylinder.

3.6.3 Cylinders subjected to external hydrostatic pressure

Cylindrical shells loaded by external hydrostatic pressure are often discretely stiffened for additional safety in the case of catastrophic events. Submarines may be regarded as an illustrative example, since in normal operation no panel buckling should occur, whereas in catastrophic situations, when the loads become excessively high, the load carrying capacity of the vessel is of interest regardless of whether the skin has buckled or not.

It is important to note that for the unstiffened cylinder under external pressure, there are only buckling and post-buckling patterns consisting of one row of buckles which extend over the whole cylinder length, Fig.3.77.

Fig.3.78 shows the load shortening curves of cylinders under external pressure in the region of small values of axial shortening. The curves were obtained for principally the same types of specimens as treated in the case of axial load: an unstiffened cylinder, a stringerstiffened, a ring-stiffened and a ring- and stringerstiffened cylinder. However, the stiffeners were doubly strong, Fig.3.71b.

The buckling pressure is strongly influenced by the arrangement of the stiffeners, contrary to the buckling loads of axially loaded cylinders. The more the one-tier pattern, which is characteristic of buckling **under external pressure**, is disturbed by the stiffeners, the higher is the buckling load.

Stringer, Fig.3.79, modify the buckling pressure only slightly by influencing the number of circumferential waves. The buckling pressure of the stringer stiffened cylinder was even slightly lower than the buckling pressure of the unstiffened one, since initial imperfections had been produced when the stringers were bonded to the skin.

Rings, Fig.3.80, curtail the longitudinal extension of the buckles and thus raise the buckling pressure considerably.

For the stringer and ringstiffened cylinder , Fig.3.81, the longitudinal extension of the buckles is curtailed, and the circumferential wave number is influenced. This leads - in our example - to a further increase of the buckling pressure.

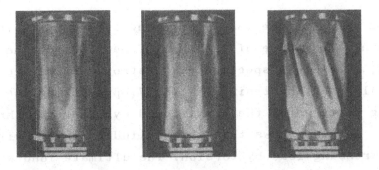

Fig.3.77 Postbuckling patterns of an unstiffened cylinder under external pressure .

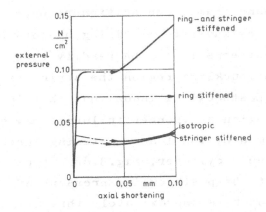

Fig.3.78 Load-shortening curves of pressure
loaded cylinders for small axial shortening .

Fig.3.79 Postbuckling patterns of the stringer
stiffened cylinder under external pressure .

Fig.3.82 exhibits the load shortening curves up to large shortening. The curve of the unstiffened cylinder has been omitted, since this specimen was destroyed before reaching large values of shortening. Fig.3.77, presenting the post-buckling patterns of the unstiffened cylinder for three pressure levels, shows that this cylinder finally withdraws from increasing load by torsion. The ultimate load was 1.3 times the buckling load.

Fig.3.79 displays the postbuckling patterns of the stringer stiffened cylinder for two pressure levels. At the higher level, right hand frame, the stiffeners become effective by preventing the cylinder from failing by torsion. From these postbuckling patterns it can be readily concluded that in the initial postbuckling region the stringers only slightly influence the postbuckling loads, but that in the advanced postbuckling region a stronger influence may be expected. This is confirmed by the load shortening curve of the stringer stiffened cylinder, Fig.3.82. It can be seen from this curve that the postbuckling pressure of the stringer stiffened cylinder is approximately three times as high as the buckling pressure.

Fig.3.80 shows the postbuckling patterns of the ringstiff-ened cylinder for two pressure levels, the first belonging to the initial and the second to the advanced-postbuckling region. Both patterns consist of three rows of buckles separated by the rings. Like the unstiffened cylinder, the ringstiffened cylinder finally fails in a torsional mode. The large influence exerted by the rings on the buckling behavior is clearly confirmed by the load shortening curve. In the advanced postbuckling region the load carried by the cylinder is not much higher than the buckling load. This relation between the buckling and the postbuckling load is

Fig.3.80 Postbuckling patterns of a ring stiffened cylinder under external pressure .

Fig.3.81 Postbuckling patterns of a stringer- and
ring stiffened cylinder under external pressure .

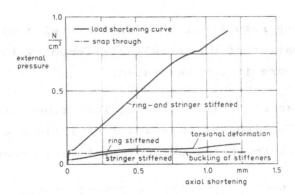

Fig.3.82 Load-shortening curves of three discretely stiffened cylinders under external hydro-
statik pressure .

similar to that found for the unstiffened cylinder. If torsional failure is possible, the postbuckling loads obviously
cannot exceed very much the buckling loads.

Fig.3.81 shows the postbuckling patterns of the stringer
and ringstiffened cylinder. For this cylinder the longitudinal wave length of the buckles is curtailed by the
rings and the circumferential wave number is influenced by
the stringers. From the fact that in the panels there are
one and two buckles alternately, it can be concluded that
the circumferential wave number suggested by the stringers
does not suit the cylinder. The postbuckling patterns for
the lower and the higher pressure level differ with respect
to the depth of the buckles only. Failure by torsion was not
observed. Corresponding to the strong influence of the stiffeners on the postbuckling pattern, the postbuckling loads are
considerably raised in the initial as well as in the advanced
postbuckling region. The diagram of the postbuckling curves,
Fig.3.82, clearly shows the strong superiority of this type
of stiffening. The test cylinder finally failed by disinte-
gration of the stiffeners rather than by loss of stability.

Summarizing the results of the experimental study on the
stability behavior of discretely stiffened cylinders under
external pressure it can be stated:

- With respect to the buckling load ring stiffeners are
  most effective. Stringers are of influence only if
  they disturb the circumferential wave number.

- With respect to the load carrying capacity in the
  postbuckling region stringers are almost as effective
  as rings, since they prevent the torsional failure
  mode. By far the highest load carrying capacity was
  found for the stringer and ringstiffened cylinder.

3.7 Isotropic cylinders under combined axial load and internal pressure

3.7.1 Experimental investigations

Now we turn to isotropic cylinders under axial load and internal pressure [26]. The theoretical buckling loads of perfect axially loaded cylinders are independent of internal pressure. However, the experimental buckling loads do depend on the pressurization, since the internal pressure weakens the deleterious influence of the initial imperfections.

In the very initial postbuckling region, axially compressed cylinders basically show similar behavior whether internal pressure is present or not. The cylinder would tend to expand under axial compression due to Poisson's ratio, and even more if any internal pressure is present. This out-ward extension is prevented at the boundaries. The constriction at the boundaries decays in wave form, Fig.3.83a, leading to an axisymmetric prebuckling deformation which increases with increasing axial load, till finally a bifurcation point is reached [27] at which stage a buckling pattern with single buckles appears on the inward wave of the axisymmetric prebuckling deformation, Fig.3.83b. In perfect cylinders the buckling pattern is periodic around the circumference of the cylinder, whereas on actual cylinders the buckling process begins with the appearance of one single buckle near one of the edges.

In the more advanced postbuckling region different behavior of unpressurized and pressurized cylinders is observed.

In unpressurized cylinders, the buckle that occurs at the boundary is the starting point for a buckling pattern that covers a great deal of the cylinder surface. One assumes

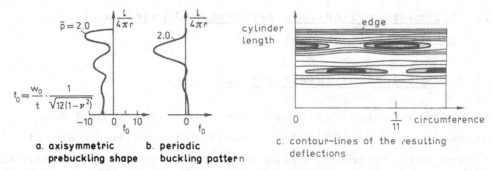

$$f_0 = \frac{w_0}{t} \cdot \frac{1}{\sqrt{12(1-\nu^2)}}$$

a. axisymmetric
prebuckling shape

b. periodic
buckling pattern

c. contour-lines of the resulting
deflections

Fig.3.83 Buckling and prebuckling deformation for internal
pressure $\bar{p} = 2.0$.

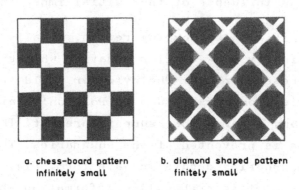

a. chess-board pattern
infinitely small

b. diamond shaped pattern
finitely small

Fig.3.84 Buckling and postbuckling pattern .

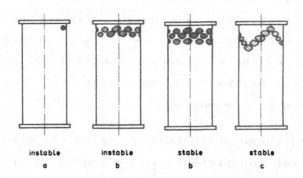

instable
a

instable
b

stable
b

stable
c

Fig.3.85 Postbuckling patterns of an axially loaded cylinder with medium internal pressure .

that at first a chess-board pattern prevails, the amplitudes of which are infinitesimal and, hence, cannot be
detected by eye, Fig.3.84a. When the buckles have become
deeper, so that they can be seen by naked eye, they have
transformed into the diamond shaped postbuckling pattern,
Fig.3.84b. This pattern does not mean a fundamental altera-
tion of the chess-board pattern; it is formed inevitably
when the inward buckles become deeper and larger while the
outward buckles shrivel together to straight narrow ridges.
These ridges move obliquely and thus yield the rhombic look.
The points of intersection are the centers of the former
outward buckles.

This rhombic postbuckling pattern is unstable and finally
transforms into the well-known stable postbuckling pattern,
consisting of two staggered rows of buckles.

The axial buckling load of unpressurized cylinders calculated for the chess-board pattern neglecting the actual
boundary conditions is approximately 10 % higher than the
buckling load calculated by properly accounting for the
boundary constriction.

In pressurized cylinders the boundary effects are stronger
than in unpressurized ones, and, hence, the postbuckling
pattern remains restricted to the boundary zone. The appearance of the first single buckle occurs at a higher
experimental buckling load than in the case of the unpressurized cylinder, and the loads carried in the postbuckling
region are also higher.

Fig.3.85 presents schematic sketches of typical postbuckling
patterns that have been observed in high-speed motion pictures of the buckling process in pressurized cylinders. One
can see that the buckling process, as in the cylinder with-

out internal pressure, begins with the snapping-in of a
single buckle near one edge, Fig.3.85a. This single buckle
is the source of a narrow postbuckling pattern, consisting
of two or three rows of buckles arranged around the circumference near this edge, Fig.3.85b. At medium values of internal pressure these tiers of buckles are unstable. Buckles are pushed towards the top and the bottom such that on
long cylinders a regular zig-zag-pattern is formed, Fig.
3.85c. In short cylinders the zig-zag-pattern is generally
not so regularly developed, since the postbuckling pattern
gets stable, before the transformation has terminated.

At high values of internal pressure the three-tier pattern
appearing immediately after buckling is stable and independent of the cylinder length, Fig.3.86.

Fig.3.87 gives a survey of the stable postbuckling patterns
of axially loaded cylinders subjected to different values
of internal pressure.

Fig.3.88 presents load shortening curves, recorded in tests
on pressurized axially loaded cylinders. The horizontal line
above the diagram is the classical buckling load, equally
valid for cylinders with and without internal pressure.
From this diagram two statements can be deduced:

- The internal pressure raises the buckling and postbuckling loads.

- With high internal pressure all postbuckling equilibrium states are stable, provided the tests are
  performed on a deformation-controlled device.

The latter statement is in agreement with the postbuckling
behavior observed with high-speed-motion pictures. At high

l = 125 mm   l = 225 mm   l = 425 mm

Fig.3.86   Pressurized axially loaded cylinders
           r = 100 mm, t = 0.254 mm, p = 3.5 N/cm^2.

p = 0   p = 1.0 N/cm^2 p = 3.50 N/cm^2

Fig.3.87   Pressurized axially loaded cylinders
           r = 100 mm, t = 0.254 mm, l = 425 mm .

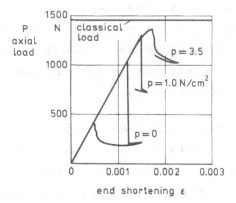

Fig.3.88   Load-shortening curve of an axially loaded cylin-
           der with different values of internal pressure .

internal pressure the initial postbuckling pattern has been
found to be stable, Fig.3.87c; at medium pressure the tiers
of buckles existing at the beginning of the buckling process turned into zig-zag-patterns, Fig.3.87b.

3.7.2 Theoretical investigations

For sufficiently long cylinders under pure axial load, neglecting the influence of boundary constriction was admissible, since this influence reduces the theoretical buckling load in the order of 10 % only, whereas the reduction
of the buckling load due to initial imperfections lies in
the order of 50 %. But for cylinders under axial load and
high internal pressure this neglection is no longer justified, since the experimental buckling loads of the imperfect cylinder may approximately reach the theoretical buckling load of the perfect cylinder, and then the influence
of boundary constriction becomes the only lowering effect.
However, postbuckling calculations for the advanced postbuckling region taking into account the boundary constriction are so difficult that till now they have not produced
useful results. So far we have been successful only in calculating postbuckling loads using the diamond shaped postbuckling pattern without boundary conditions. The essential
result of these computations is the influence of the internal pressure on v.Kármán's load.

For the calculation of v.Kármán's load it is most important
to know the aspect ratio of the first buckle. On the motion
pictures it seems to be discernible that the buckles snapping-in are square at small internal pressure and are longer
in circumferential than in axial direction at high internal
pressure.

The perfect pressurized cylinder would buckle with an axisymmetric pattern. From this, one may deduce that the axisymmetric imperfections are amplified most strongly. Koiter [17] stated, that in cylinders with axisymmetric imperfections the axial wave length of the periodic buckling pattern is twice that of the imperfection pattern. If one combines this statement with the experimental result that in cylinders with small internal pressure, the buckles snapping-in are square, one is led to the circumferential wave number n = 18 , which is the same as for the unpressurized perfect cylinder.

We performed principally the same computations as carried out before for the unpressurized cylinder, i.e. we computed the smallest potential energy for different circumferential wave numbers and aspects ratios of the buckles. In this case the equilibrium states of smallest shortening and smallest potential energy do not coincide, since the radial pressure contributes to the potential energy.

Fig.3.89 presents the results of these computations for a cylinder with small internal pressure:

- On the curve d = 1 , emphasized in the diagram by a vertically hatched plane, there is again a distinct minimum, indicating the end of the intrinsic buckling process. However, compared with unpressurized cylinders, this minimum is situated at a higher circumferential wave number and corresponds to a higher axial load.

- It is remarkable that for the cylinder with internal pressure an absolute minimum of the potential energy exists. For small internal pressure this minimum is

connected with such a large axial wave length, that
it could not be attained by our test cylinders.

● The absolute minima of potential energy obtained in
our calculations approach v.Kármán's load. The existence of these minima does not imply that the corresponding equilibrium states are stable, since the
computations were restricted to diamond-shaped postbuckling patterns; at medium internal pressure the
cylinder escapes into a zig-zag-pattern, not contained in the analysis.

In Fig.3.90 the potential energy has been plotted as a function of the circumferential wave number for the aspect ratio
d = 1 and for various values of internal pressure. It can
be seen that the minima are clearly marked and with increasing internal pressure move to higher wave numbers.

We compared v.Kármán's load with our experimental buckling
loads for cylinders with radius-to-wall-thickness ratio
$r/t = 100/0.254 = 394$. These are our standard cylinders
for which the motion pictures were taken and the calculations were performed. We found that all our experimental
buckling loads of pressurized as well as of unpressurized
cylinders lay above v.Kármán's load. In order to compare
v.Kármán's load with experimental buckling loads of other
authors we took from a report of Weingarten, Seide and
Morgan [28] the experimental lower limit of the scatter
region of buckling loads for $r/t = 500$, since this ratio
corresponds well to our specimens. The comparison, in non-
dimensional terms, is presented in Fig.3.91. For small internal pressure the agreement is good. For high internal
pressure the computations yield no reduction of the classical buckling load, whereas the measured buckling loads

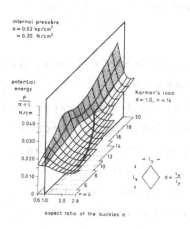

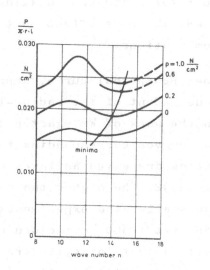

Fig.3.89 Smallest values of potential energy .

Fig.3.90 Minimum potential energy for buckles with aspect ratio d = 1 .

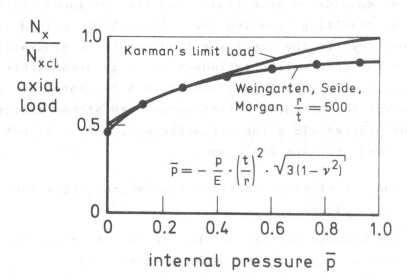

Fig.3.91 Calculated and measured lower limits for the axial loads of isotropic cylinders .

are lower. This discrepancy may be explained by the fact
that for cylinders buckling at loads close to the classical
load, the edge effect must no longer be neglected in the
computations.

The theoretical curve, representing v.Kármán's loads, is
independent of the radius-to-thickness ratio r/t , whereas
in the work of Weingarten et.al. [28] different experimental curves corresponding to different radius to thickness
ratios are given as lower bounds on the buckling loads,
Fig.3.92. The higher the radius-to-thickness ratio r/t
the smaller the experimental buckling loads. As the curve
for $r/t = 500$ coincided well with v.Kármán's load, it
follows that for cylinders with $r/t > 500$ there might
be experimental buckling loads lower than v.Kármán's load.
In order to explain why for these very thin-walled cylinders the experimental buckling loads may be smaller than
the v.Kármán's load, we recall that the v.Kármán's load
is to be considered as a limit load for the local rather
than the resulting buckling load. If v.Kármán's load is
attained locally, the resulting load may be considerably
smaller for thin-walled cylinders. This is demonstrated
in Fig.3.93. In this figure you can see for three different radius-to-thickness ratios normalized stress distri-
butions plotted along the circumference. For each r/t
ratio three stresses are shown:

> the actual stress distribution which limits the
> shadowed area,

> the average stress, which corresponds to the re-
> sulting experimental buckling load, and

> the stress corresponding to v.Kármán's load.

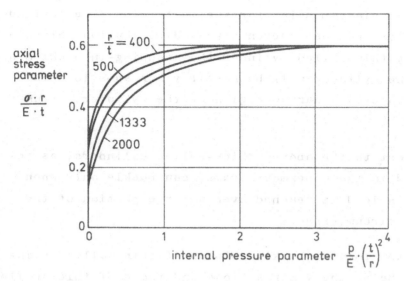

Fig.3.92 Axial buckling stress with internal pressure .

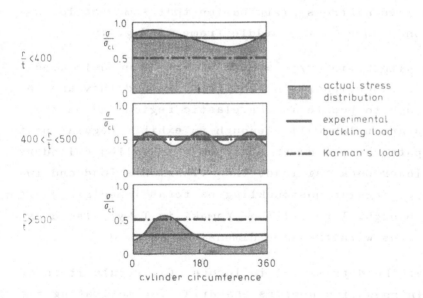

Fig.3.93 Comparison of v.Kármán's load with actual stress .

The lowest diagram holds for the most thin-walled cylinder. It exhibits the most uneven stress distribution. Since on that very thin-walled cylinder the snapping-in buckles are small, the cylinder will buckle if v.Kármán's load is reached in even a narrow region of the cylinder circumference.

In contrast to the above, thick-walled cylinders, as represented in the uppermost frame, can buckle only when v.Kármán's load is reached over a large portion of the cylinder circumference.

If the experimental buckling loads of thin-walled cylinders lie below the v.Kármán load and those of thick-walled cylinder lie above it, then it follows that there is a radius-to-thickness ratio between the two, where the experimental buckling load agrees with the v.Kármán load. In such a case we have a stress distribution that somewhat looks like the one shown in the middle frame of Fig.3.93.

Cylinders with ratio r/t about 500 are frequently used in buckling experiments since on the one hand they are thin-walled enough to buckle in the elastic region and on the other hand are thick-walled enough to exhibit regular post-buckling patterns. We never observed thin-walled cylinders that had lower buckling loads than v.Kármán's load and in the meantime, regular postbuckling patterns immediately after the snap-through. I recall that Yamaki in Tokyo also uses test specimens with the same dimensions as we do.

Pointing at the detrimental influence of irregularities of the load introduction appears essential for motivating the engineer to be as careful as possible in designing the respective details.

3.8 Isotropic cylinders under combined axial load and external pressure

3.8.1 Buckling investigations

For cylinders under axial load and external pressure, both
loads contribute to the buckling of the cylinder. Fig.3.94
shows the theoretical interaction curve of a perfect cylinder with length parameter $\bar{l} = 36.8$ and test boundary conditions. The presentation is non-dimensional. The cylinder
buckles under pure external pressure when the value $p/p_{cr} = 1$
on the abscissa is reached, or similarly under pure axial
load for the value $N_x/N_{xcr} = 1$ on the ordinate. One can
observe that small external pressures $p/p_{cr} < 0.5$ have
only small influence on the axial buckling load; the external pressure starts to reduce the axial buckling load considerably, not before $p/p_{cr} < 0.5$ has been attained.

3.8.2 Postbuckling investigations

Fig.3.95 presents the postbuckling patterns, which were
stable immediately after the snap-through on a cylinder
with length parameter $\bar{l} = 58.4$ subjected to axial load
and external pressure. The series starts with the cylinder
under pure axial load, left hand frame, and finishes with
the cylinder under pure external pressure, right hand frame.
From this figure it can be observed that the two-tier pattern occurring under pure axial compression is maintained
as long as the external pressure does not exceed 59 % of
the theoretical buckling pressure. The pressure has just
the effect of making the buckles more elongated in the
axial direction. On the other end of the series the one-
tier pattern, typical for buckling under external pressure
is shown. With increasing axial load this pattern is main-

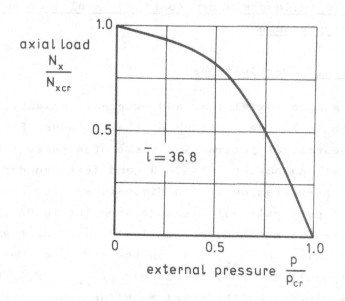

axial load $\dfrac{N_x}{N_{xcr}}$

external pressure $\dfrac{p}{p_{cr}}$

$\bar{l} = 36.8$

Fig.3.94   Interaction curve .

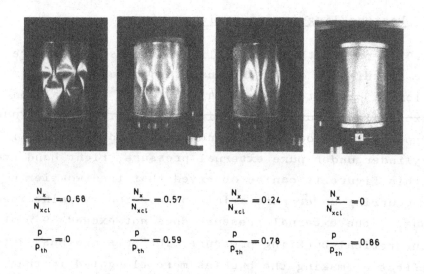

$\dfrac{N_x}{N_{xcl}} = 0.68$ $\dfrac{N_x}{N_{xcl}} = 0.57$ $\dfrac{N_x}{N_{xcl}} = 0.24$ $\dfrac{N_x}{N_{xcl}} = 0$

$\dfrac{p}{p_{th}} = 0$ $\dfrac{p}{p_{th}} = 0.59$ $\dfrac{p}{p_{th}} = 0.78$ $\dfrac{p}{p_{th}} = 0.86$

Fig.3.95   Postbuckling patterns of cylinder subjected
to axial load and external pressure .

tained as long as the axial load remains below 24 % of the theoretical axial buckling load. The axial load has just the effect of pushing the buckles to the middle of the cylinder, giving them an appearance that somewhat reminds of the diamond shape. We did not determine exactly the load combination where transition from the one-tier pattern to the two-tier pattern occurs.

The fact that there is no smooth transition from the one-tier to the two-tier postbuckling pattern is reflected in the theoretical interaction curve, Fig.3.94, by the distinct change of the slope. It may be pointed out that the course of this curve and especially the region where the slope changes most rapidly, does not suit to the cylinder shown in Fig.3.95, since the length parameters $\bar{l}$ are different and furthermore the experimental buckling loads are always reduced by initial imperfections, this reduction being greater for axial compression than for external pressure.

In Figs.3.96 and 3.97 load-shortening curves for cylinders subjected to combined loads are shown. The curves of Fig. 3.96 were compiled for constant external pressure and varying axial load. You can see three load shortening curves. The first one corresponds to $p = 0$, i.e. to pure axial compression; it is well known from chapter 3.2. The second postbuckling curve was recorded at the external pressure $p = 0.04 \ N/cm^2$ which is 39 % of the theoretical buckling pressure. This pressure reduces the buckling load only little, but the postbuckling loads are reduced to nearly half the values of the cylinder under pure axial compression. From the preceding discussion of the postbuckling patterns shown in Fig.3.95 we know that for the load combination with an external pressure less than 59 % of the theoretical buckling load the two-tier postbuckling pattern prevails.

184

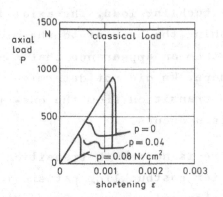

Fig.3.96   Load-shortening curves of axially loaded cylinders
with constant values of external pressure .

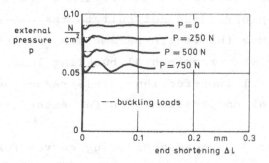

Fig.3.97   Load-shortening curves of pressure loaded cylinder
with constant values of axial load .

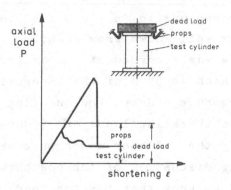

Fig. 3.98   Partition of load in dead weight test assembly .

The third postbuckling curve of Fig.3.96 pertains to the external pressure p = 0.08 N/cm^2 which is 78 % of the theoretical and 90 % of the experimental buckling pressure. There is a strong reduction of the buckling load and a strong influence on the postbuckling load carried just after the snap through. From the discussion of the postbuckling patterns presented in Fig.3.95 we know that at the load combination with an external pressure which is 78 % of the theoretical buckling pressure a one-tier postbuckling pattern appears after the snap-through. From the fact that now we have a postbuckling pattern which is different from that of the cylinder under pure axial load it follows that the postbuckling behavior is also changed. The slope of this postbuckling curve is steeper than for the other curves. If the axial load is totally removed, the cylinder remains in a postbuckling equilibrium state, since the constant external pressure is higher than the minimum postbuckling pressure of the cylinder.

The curves presented in Fig.3.97 are obtained for different constant axial loads and varying external pressure. The test assembly was rather difficult, for the constant axial load was exerted by weights which had to be securely intercepted upon buckling, so that our measuring apparatus was not damaged. One can see from the diagram that the buckling pressure is the smaller, the higher the constant axial load. The decrease of buckling pressure is almost linear with the increasing axial load. In this test assembly it was not possible to remove the buckles by reducing the external pressure. This is a similar situation as the one discussed for the lowermost postbuckling curve of Fig. 3.96. But things are somewhat more complicated here because the weights are partly carried by the intercepting props. The equilibrium state of the buckled cylinder is illustrated

in Fig.3.100. The test cylinder is in the advanced postbuckling region, and its axial postbuckling load is smaller
than the dead axial load exerted by the weights. The weights
are carried only partly by the buckled cylinder and partly
by the props. The cylinder is so much weakened by the buckling that it can no longer lift the weights by itself.

Theoretical postbuckling investigations of the cylinder
under combined axial load and external pressure are not
known to us, even though there would not be any difficulties, because in this loading case the buckles are large.
This is in contrast to the situation for cylinders under
combined axial load and internal pressure, where the buckles are small and an exact postbuckling calculation would
be very difficult.

3.9   Axially loaded cylinders buckling in the plastic region

3.9.1   Preliminary remarks

In this section only test results will be treated. Postbuckling calculations in the plastic region which agree
with test results are not known to us. At any rate, in
these calculations, one must not only know the yield limit
but also the behavior of material in the region beyond the
elastic limit including work hardening.

In the following we will report on plastic buckling up into the very advanced postbuckling region. Those tests are
of practical importance in two cases:

- First, catastrophic failure, i.e. investigations in
  order to obtain an explanation for the collapse of
  a structure.
- Second, energy absorption, for instance experiments
  during the development of a Safe Auto.

3.9.2 Circular cylindrical tubes with low yield limit

In Fig.3.99 you see the fragments of a spherical container.
This container was originally supported by 16 columns, which
were circular tubes. Due to insufficient care in assembling
the columns to the foundation, there was no uniform distribution of the load on the columns. Those responsible for
erecting the structure originally were of the opinion that
the loads would eventually reach a uniform state as each
column could only be loaded up to its yield limit, and then
would carry a constant load.

After the structure had collapsed two explanations for the
failure were considered:

- In the first one it was assumed that the collapse was
  initiated by fracture of the container, Fig.3.99a, due
  to use of a material that was not capable of sufficient plastic deformation for avoiding high local
  stresses.

- The second explanation was based on the assumption
  that the supporting tubes buckled locally, Fig.3.99b,
  and that after buckling the load carried by a tube
  decreased. This would have resulted in a sequence of
  tube bucklings, the first supporting tube having been
  overloaded and having buckled would have been followed by the second and so on.

The tubes were so thick-walled that they could buckle only
in the plastic region. In order to determine the buckling
loads of the tubes and their behavior in the postbuckling
region we produced test specimens of nearly the same radius-
to-thickness ratio as the actual tubes. These specimens were
made out of steel St 35 and St 52 respectively.

a. Spherical tank

b. Supporting tube

Fig.3.99 Collapsed container.

The dimensions were: radius $r = 48.5$ mm , wall-thickness $t = 1.0$ and 1.8 mm , length $l = 80$, 100 and 161 mm , respectively. The circular shape of the cylinder edges was maintained by rigid end plates; but these end plates were not fastened to the cylinder wall.

Fig.3.100 presents three cylinders with plastic postbuckling deformations, cut open and photographed after the buckling test. The centrally loaded cylinder **a** has buckled only at one end. All centrally loaded cylinders of this test series exhibited the same buckling and postbuckling behavior. The explanation for this behavior is simple. The formation of axisymmetric buckles is favoured by the axisymmetric prebuckling deformations which are due to edge constriction. Hence, the cylinder tends to buckle near the edges. If buckling has started at one edge, this edge zone is weakened, offering little resistance to further deformation, in contrast to the opposite unbuckled edge. Further increase in axial stress, which could cause the unbuckled edge to buckle, does not take place.

The postbuckling patterns **b** and **c** of Fig.3.100 are produced by eccentric axial loading. Here also the buckling process began at one edge. However, the deformations cover a larger portion of the cylinder length, since the cylinder axis is bent.

Fig.3.101 exhibits the buckling loads as function of the load eccentricity. One sees immediately that the centrally loaded cylinder buckles as soon as the yield limit is reached. For the eccentrically loaded cylinder with medium eccentricity $e/r = 0.5$ the same statement is valid. You may easily verify that by some short calculations. In the case of high eccentricity $e/r = 1$ the stress calculation

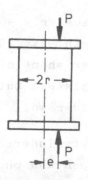

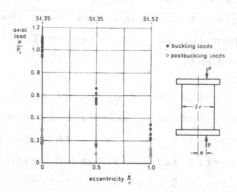

Fig.3.100 Postbuckling patterns of plastically buckled cylinders

Fig.3.101 Plastic buckling loads of axially loaded cylinders

would be rather complicated, since the cylinder wall lifted
off from the end plates and consequently the strain distribution was no longer linear over the cross section of the
tube.

Fig.3.102 shows the load shortening curve of the cylinder
under central axial loading. In this buckling process the
plastic postbuckling pattern of Fig.3.100 a is produced.
In the prebuckling region the load and shortening increase
proportionally. We had to stop the shortening process as we
reached the end of the working capacity of our test set up.
In decreasing the shortening the load shortening curve moves
parallel to the rising straight line in the elastic region.

Fig.3.103 shows the load shortening curves for eccentrically
loaded tubes. During these tests the postbuckling patterns
shown in Fig.3.100 b and c were formed. As mentioned in discussing Fig.3.101, the buckling loads under eccentric loading are essentially smaller than those under central loading,
since the cylinder buckles when the yield stress is attained
at the most stressed place on the circumference.

In order to complete the picture, we also tried to investigate the influence of non-uniformity of stress distribution
on the buckling and postbuckling behavior. We introduced
washers of up to 1 mm thickness under the cylinder edges,
Fig.3.104a. However the result was disappointing. The washers
impressed themselves in the cylinder wall without influencing
the loading behavior in the least, Fig.3.104b.

This was the last of the experiments performed in order to
find out whether the spherical container failed due to the
buckling of the supporting tubes. The question could not be
categorically answered by the tests. The fact that the load

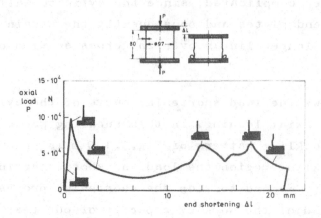

Fig. 3.102 **Load-shortening of a centrally loaded plastically buckling cylinder .**

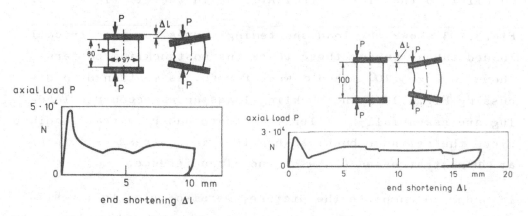

Fig. 3.103 **Load-shortening curves of eccentrally loaded
plastically buckling cylinders .**

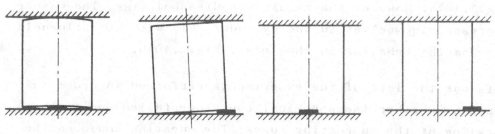

a. before buckling   **b. after buckling**

Fig. 3.104 **Disturbed buckling process .**

carried by the column drops after buckling supports this
hypothesis; but that this load drop is connected with a
large amount of shortening speaks against it.

3.9.3 Circular cylindrical tubes with high yield limit

The Volkswagen plant is developing a Safe Auto and in the
course of these investigations tubes made of steel St 35
with radius r = 38.5 mm , wall-thickness t = 1.0 mm and
length l = 1000 mm , were compressed [29]. The material
was highly work hardened by cold rolling. The ultimate
stress exceeded 500 N/mm^2 . The yield stress was only a
little bit lower.

Fig.3.105 shows a plastically buckled tube. One sees that
the postbuckling pattern is not axisymmetric, but exhibits
n = 4 waves in circumferential direction. It consists of
a number of staggered rows of diamond-shaped buckles compressed together in the axial direction until their flanks
lie one upon another. In the actual test one row of buckles
after another formed and got compressed in the axial direction yielding the above picture. If the cylinder would have
been compressed further the postbuckling pattern would even
show some more rows of buckles.

In Fig.3.106 the load shortening curve is shown. We see that
the tube was compressed to 250 mm, i.e. to a quarter of its
original length.

The buckling load, that is the load at which the first row
of buckles occurs is $9 \cdot 10^4$ N : this corresponds to the
stress σ = 377 N/mm^2 . This stress is higher than the buckling stress of the test specimens, the buckling and postbuckling behavior of which has been discussed in section
3.9.2 . The difference in the buckling stresses is probably

Fig.3.105   Postbuckling pattern of an axially loaded
            plastically buckled tube .

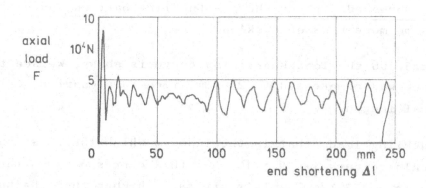

Fig.3.106   Load-shortening curve of an axially loaded
            plastically buckled tube .

the cause of the phenomenon that the tubes in Fig.3.100
and 3.105 have different postbuckling patterns inspite of
their nearly equal radius-to-wall-thickness ratio.

On the appearance of the first row of buckles the load
carried by the cylinder drops sharply. With further increase in the shortening the load begins to rise, till at
$4 \cdot 10^4$ N the second row of buckles appears. Again the
load falls at this stage, and with further compression
begins to rise. Successive rows of buckles follow each
other and with each formation of a new row there is a
corresponding drop in the load and subsequent increase.
Thus a zig-zag load-deformation curve is produced which
oscillates around a mean value $P = 3.5 \ 10^4$ N . The
buckling load is not attained any more in the postbuckling
region.

This shock absorber has the advantage that the force, apart
from the initial maximum value, is nearly constant and independent of the shortening; it has the disadvantage that
it requires extra weight and that it might not avoid damage
right at the beginning when the maximum value of the force
occurs.

These disadvantages are decreased by an improved design
discussed in the following.

3.9.4 Channel section with closing plate

Instead of dealing with a circular tube section, the improved design has a built-up cross section consisting of
a U-shaped channel and a closing straight plate, Fig.3.107.

The channel section is formed out of a 1.75 mm thickness
sheet of Soldur quality 40, and the straight plate is 0.75 mm
thick and is made from steel St 1023. The straight plate and

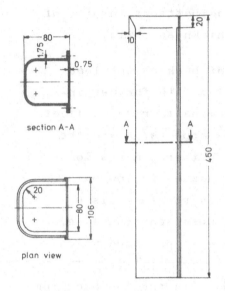

Fig.3.107 U-section with
closing plate .

Fig.3.108 Postbuckling pattern of the U-section with
closing plate .

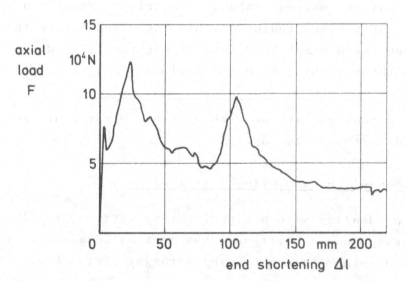

Fig.3.109 Load-shortening curve of the U-section with
closing plate .

the U-profile are spot-welded together. This section has
the advantage that it can be used not only as a shock absorber but also as a beam for the engine mount. For the
buckling test this built-up column was welded at one end
to an end plate. The other end was chamfered, in order to
reduce the peak stress that would otherwise occur at the
instance of buckling as has been observed with the cylindrical tube.

Fig.3.108 shows the buckled structure. One sees clearly
that as a result of plastic buckling a lot of the shock
energy is absorbed. Fig.3.109 shows the postbuckling diagram. The highest peak of stress is about double the mean
value. It does not occur at the first instance of shock
but only after 20 mm of motion when the entire chamfering
is levelled and folded over.

The essential criterion for evaluating the different cross
sections is their capacity for energy absorption. This is
the area under the load shortening curve. To get index
values for comparison it is related to the cross sectional
area of the column.

The section shown in Fig.3.107 does not provide the optimum
solution for the problem of the best possible energy absorber
for a Safe Auto design. We have chosen an arbitrary example
from a test-series to show an important practical application of postbuckling behavior in the plastic regime.

198

Literature

[1]   ESSLINGER, M.E.   Buckling and Postbuckling Behavior
       GEIER, B.M.   of Thin-Walled Circular Cylinders.

                           Proc. of the Internat. Coll. on
                           Progress of Shell Structures.
                           Session IV, Madrid, Sept.-Oct.
                           (1969), pp. 68

[2]   MEYER-PIENING,H.R. Theoretische und experimentelle
                           Untersuchung des Nachbeulverhaltens
                           dünnwandiger Kreiszylinder mit ein-
                           gespannten Rändern unter Außendruck-
                           belastung.

                           DLR-FB 70-09, Jan. (1970), 89 S.

[3]   THIELEMANN, W.F.   Problems of Buckling of Thin-Cylin-
       ESSLINGER, M.E.   drical Shells.
       GEIER, B.M.

                           DFVLR Bericht F 63-10 (1963)
                           (unpublished report)

[4]   KARMAN, Th.   The Buckling of Thin Cylindrical
       TSIEN, H.S.   Shells under Axial Compression.

                           J. Aeron. Sci., Vol.8 (1941),
                           pp.303-312

[5]   ESSLINGER, M.E.   On the Buckling and Postbuckling of
       GEIER, B.M.   Thin-Walled Circular Cylinders.

                           RILEM, Internat. Symp., Buenos-
                           Aires (Argentine), 13-18. Sept.
                           (1971), pp.97-124

[6]   ALMROTH, B.O.   Postbuckling Behavior of Axially
                           Compressed Circular Cylinders.

                           AIAA J., Vol.1 (1963), p. 630

[7]   HOFF, N.J.   Postbuckling Equilibrium of Axially
       MADSEN, W.A.   Compressed Circular Cylindrical
       MAYERS, J.   Shells.

                         AIAA, Vol.4 (1966), pp.126-133

[8]   MADSEN, W.A.   The Snap-Through and Postbuckling
       HOFF, N.J.   Equilibrium Behavior of Circular
                           Cylindrical Shells under Axial Load.

                         Dept.of Aeron.and Astron.,Stanford
                         Univ., SUDAER No.227 (1965)

[9]   ALMROTH, B.O.   An Experimental Study of the Buck-
      HOLMES, A.M.C.   ling of Cylinders under Axial Com-
      BRUSH, D.O.   pression.

                         Experimental Mechanics (1964), p.263

[10] TENNYSON, R.C.   An Experimental Investigation of the
                         Buckling of Circular Cylindrical
                         Shells in Axial Compression using
                         the Photoelastic Technique.

                         Univ. of Toronto, Inst. for Aero-
                         space Studies, Rep. No.102 (1964)

[11] THIELEMANN, W.F.   On the Postbuckling Behavior of Thin
                         Cylindrical Shells.
                         "Collected Papers on Instability of
                         Shells Structures - 1962".

                         NASA TN D-1510, Washington D.C.,
                         (1962), p.203

[12] THIELEMANN, W.F.   New Developments in the Nonlinear
                         Theories of Buckling of Thin Cylin-
                         drical Shells.

                         Proc. of the Durand Centennial Con-
                         ference, Pergamon Press Inc., New
                         York (1960), p.76

[13] YOSHIMURA, Y.   On the Mechanism of Buckling of a
                         Circular Cylindrical Shell under
                         Axial Compression.

                         NACA TM 1390, July (1955)

[14] KOITER, W.T.   On the Stability Elastic Equilibrium.

                         NASA TT F-10.833

[15] KEMPER, J.   Postbuckling of an Axially Compressed
      CHEN, Y.N.   Oval Cylindrical Shell.

                         AFOSR Scientific Rep. No. 68-1994,
                         Polytechnic Inst. of Brooklyn,
                         PIBAL Rep. No. 68-31 (1968)

[16] THIELEMANN, W.F.   On the Postbuckling Behavior of Thin-
      ESSLINGER, M.E.   Walled, Axially Compressed Circular
                         Cylinders of Finite Length.

                         Proc. of the Symp. on the Theory of
                         Shells, (1967), Univ.of Houston
                         (Texas), pp.433-479

[17] KOITER, W.T.   The Effects of Axisymmetric Imperfection on the Buckling of Cylindri-
cal Shells under Axial Compression.

Lockheed Missiles & Space Comp.,
Sunnyvale, Calif., Techn. Rep.
6-90-93-86, (1963) Aug., 21 pp.

[18] ARBOCZ, J.   The Effect of General Imperfections
on the Buckling of Cylindrical
Shells.

PhD Thesis, California Inst. of
Technology, Pasadena, Calif. (1968)

[19] YAMAKI, N.   Experiments on the Postbuckling
Behavior of Circular Cylindrical
Shells under Torsion.

IUTAM Symp. on Buckling of Structures, Harvard Univ., June 17-21
(1974)

[20] BUDIANSKY, B.   Postbuckling Behavior of Cylinders
in Torsion.
Theory of Thin Shells.

Ed. by F.I. Niordson, Springer
(1969), pp.212-233

[21] DONNELL, L.H.   Stability of Thin-Walled Tubes
under Torsion.

NACA Rep. No. 479 (1933)

[22] LUNDQUIST, E.E.   Strength Tests on Thin-Walled
Duralumin Cylinders in Torsion.

NACA TN No. 427 (1932)

[23] GARKISCH, H.D.   Experimentelle Untersuchung des
Beulverhaltens von Kreiszylinder-
schalen mit exzentrischen Längs-
versteifungen.

DLR-FB 67-75 (1967)

[24] ESSLINGER, M.   Beulen und Nachbeulen exzentrisch
versteifter dünnwandiger Kreiszylin-
der unter axialsymmetrischer Bela-
stung.

DLR-FB 70-48 (1970)

[25] ESSLINGER, M.   Buckling and Postbuckling Behavior
      GEIER, B.   of Discretely Stiffened Thin-Walled
                          Circular Cylinders.

                          Z. Flugwiss., Vol.18 (1970),p.246-253

[26] ESSLINGER, M.E.   Calculated Postbuckling Loads as
      GEIER, B.M.   Lower Limits for the Buckling Loads
                          of Thin-Walled Circular Cylinders.

                          IUTAM Symp. on Buckling of Struc-
                          tures, Harvard Univ., June 17-21
                          (1974)

[27] FISCHER, G.   Über den Einfluß der gelenkigen Lage-
                          rung auf die Stabilität dünnwandiger
                          Kreiszylinderschalen unter Axiallast
                          und Innendruck.

                          Z. Flugwiss., Vol.11 (1963), p.111-119

[28] WEINGARTEN, V.I.   Elastic Stability of Thin-Walled Cyl-
      SEIDE, P.   indrical and Conical Shells under
      MORGAN, E.J.   Combined Internal Pressure and Axial
                          Compression.

                          AIAA J., Vol.3 (1965) pp.1118-1125

[29] DANCKERT, H.   VW-Forschungsbericht Nr. F 4-K-73
      MASKE   (1973)
                          (unpublished report)

4. SPHERICAL SHELL

4.1 Initial imperfection sensitivity

In striving to fully understand the mechanism of initial
imperfection sensitivity, let us briefly recall the stability behavior of flat plates and circular cylinders. In
each of these structures the relationship between the buckling and the postbuckling loads was definitively attribut-
able to the imperfection sensitivity of the structure. Thus
in the case of flat plates the postbuckling loads are higher
than the buckling load and, hence, we conclude that flat
plates are imperfection insensitive. In the case of cylinders under external pressure, the postbuckling loads are
slightly below the buckling load and so, cylinders under
external pressure are imperfection sensitive to a small
degree. With cylinders under axial loading the postbuckling
loads are quite below the buckling load, hence, axially
loaded cylinders are imperfection sensitive to a very high
degree.

Now when we turn to spherical shells we find that we have
to examine the concept of imperfection sensitivity more
deeply because the above simple relationship between the
postbuckling behavior and the lowering of buckling loads
by initial imperfections does no longer hold. With complete spheres under external pressure the postbuckling
loads are considerably below the theoretical buckling loads
of the perfect structure as it is the case with axially
loaded cylinders; furthermore there is also considerable

scatter of experimental buckling loads. However in contrast to the experience with cylinders, it was soon possible, after some initial difficulties, to manufacture thin-walled spheres which attained as much as 80 % of the theoretical buckling load [1].

From the different experiences with cylinders and spheres there arises a set of thought-provoking questions: Why are the experimental buckling loads of thin-walled spherical shells so high in spite of their low postbuckling loads? Should one classify spherical shells as imperfection sensitive with regard to the buckling loads? What exactly is imperfection sensitivity?

● The first question was: Why are the experimental buckling loads of spheres so high in spite of their low postbuckling loads? Obviously low postbuckling loads are only a necessary, but not a sufficient condition for the lowering of the buckling loads due to initial imperfections. Further necessary conditions are that these low postbuckling loads can be obtained with small normal deflections and also that initial imperfections are actually present. We believe that it is easier to manufacture relatively imperfection-free thin-walled spheres than thin-walled cylinders, since for thin-walled cylinders the connection between the cylinder wall and the bulkheads turns out to be a source of inevitable disturbances with regard to the shape of the cylinder and with regard to the load introduction. With closed shells, i.e. with the complete spherical shell, these disturbances are non-existent and, hence one can suppose that the experimental buckling loads of spheres would be closer to the theoretical values than those of cylinders.

- The second question was: Should one classify spherical shells as imperfection sensitive? Our opinion is: Yes, the spherical shell is imperfection sensitive, since the buckling load can be strongly lowered by the initial imperfections; this statement holds in spite of the fact that this imperfection sensitivity is not damaging, as it is possible to manufacture essentially imperfection-free spherical shells. In contrast to this with thin-walled cylinders deleterious imperfections can not be avoided so far.

- The third question was: What exactly is imperfection sensitivity? A simple answer would be: A structure is imperfection sensitive in relation to some character-istic feature, if this characteristic c a n be af-fected severely by small initial imperfections. But this answer still leaves the question open, for what reasons a certain shell turns out to be imperfection sensitive with regard to its buckling load.

I place these thoughts on imperfection sensitivity at the beginning of the discussion of spherical shells, so that you can follow the further report with a critical eye be-ing aware that we strive for a clear understanding of imper-fection sensitivity.

4.2 Complete sphere

4.2.1 Test specimens

Fig.4.1 shows one of the test specimens used by Berke and Carlson [2] for their buckling and postbuckling experiments on complete spherical shells. Of course, the authors chose the best sample for this figure. Nevertheless we are astonished

that this brilliant sphere, which because of its brightness
should exhibit even the slightest initial imperfections,
appears in such a perfect beauty. Obviously a production
method was used that practically excludes initial imperfections.

The sphere was made by electroplating on a wax mandrel. This
method of forming shell specimens was introduced by Thompson
[3] in 1960 and has since then been increasingly used in
experiments for shell buckling.

The sphere shown in Fig.4.1 is made of nickel and has a
radius of 108 mm and a wall-thickness of 0.05 mm. The variation of the wall-thickness within the area in which the first
buckle has jumped in, was less than 5 %. Characteristics of
the material are Young's modulus $1.97 \cdot 10^5$ N/mm^2 and the
modulus of rigidity $G = 1.02 \cdot 10^5$ N/mm^2 .

4.2.2 Experimental equipment and procedures

It is customary and illustrative to represent the buckling
and postbuckling processes in the form of load-deformation
curves. For longitudinally compressed plates and axially
compressed cylinders one plots load-shortening curves, because the shortening is theoretically particularly signif-
icant and can easily be measured. In the postbuckling investigations of cylinders under external pressure we have
kept the shortening as a deformation quantity, because it
can easily be measured. However, in this loading case it
gives no useful information. It would have been more reasonable to introduce the volume increment as the deformation
quantity, since the product of the volume increment and the
external pressure represents the potential energy of the external forces. In postbuckling investigations of complete

206

R = 108 mm $\frac{R}{t}$ = 2160
t = 0,05 mm

Fig.4.1   Test specimen

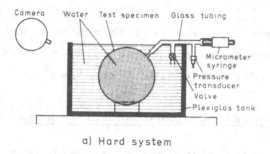

a) Hard system

a. Hard system

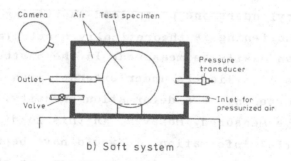

b) Soft system

b. Soft system

Fig.4.2   Loading device .

spherical shells the volume increment is always used as the deformation quantity.

In the postbuckling investigations on axially loaded cylinders we distinguished between shortening-controlled and load-controlled tests. In the load-controlled tests, the upper cylinder bulkhead had to be intercepted when the snap through began, lest the cylinder completely collapsed.

There are corresponding procedures for the postbuckling tests with spherical shells. The tests can be performed in a deformation-controlled manner by steadily reducing the volume in the prebuckling region and by keeping it constant during the buckling process. In the postbuckling region one can plot pressure-volume curves by increasing or reducing the volume. Fig.4.2a is a schematic sketch of the test facility. You see the test specimen in a plexiglass tank which is open at the top. Both the tank and the sphere are filled with water. A glass tubing is connected to the sphere. A junction pipe between the glass tubing and the water in the tank which can be closed by a valve, provides the zero adjustment. The volume in the test specimen can be enlarged or reduced by means of the micrometer syringe. A pressure transducer provides the measurement of the pressure variation. This test installation is called hard-loading system in technical language.

In the case of the soft loading system the pressure was controlled and kept constant during the buckling process. The plexiglass tank was closed and pressurized with air, while the specimen was vented to the atmosphere, Fig.4.2b. A pressure transducer was used to measure the pressure in the tank.

In the following the pressure values will be presented in relation to the classical buckling pressure

$$P_{cl} = \frac{2E}{\sqrt{3(1-\gamma^2)}} \cdot \left(\frac{t}{R}\right)^2,$$

which is the buckling pressure of the perfect shell.

4.2.3 Experimental results

Two types of tests were conducted in the hard system. In
the first the wax mandrel was not removed from the nickel
sphere. Electroplating occurs at a slightly elevated temperature and, when specimen and mandrel are cooled to room
temperature, a gap develops between the nickel sphere and
the wax mandrel. Spheres were buckled in this configuration
in order to obtain high-speed, close-up motion pictures of
the initiation of the buckles. Buckling tests in this configuration are repeatable because the extent of buckling
is controlled. The buckling behavior for rather imperfect
spheres, which buckled at a pressure less than 50 % of the
theoretical value was different from that of more perfect
shells which buckled at a pressure greater than 50 % of the
theoretical value.

At low buckling pressures it was generally found that buckling starts with the snapping-in of a single dimple whose
size depended on a gap dimension. Examination of the specimen revealed that the dimple was at a flaw. If the test was
continued, additional dimples were observed to pop in, usually singly and at higher pressures. The subsequent dimples
formed at flaws which appeared less severe than the flaw at
the first dimple.

At higher buckling pressures the shell was more highly
stressed and the transition to the buckled state occurred
with the formation of a large number of dimples. The

transition was very rapid, and it was not possible to
detect a dimpling sequence by visual observation, i.e. the
dimples appeared to form simultaneously. An example of the
resulting pattern is shown in Fig.4.3.

The buckling process at a high buckling pressure $(0.9\ p_{cl})$
was filmed using a high-speed camera. The location of the
first of many small dimples, which seemingly develop in-
stantaneously, could be determined by analysing the high-
speed motion pictures taken of a large suspected area before a close-up of the first dimple was attempted. Despite
this technique, several tests were required to obtain successful close-ups such as the one shown on Fig.4.4. This
was necessary because the first buckle did not always occur
at exactly the same location, even if the buckling pressure
was the same in successive tests carried out with the same
sphere. This suggests the absence of plastic deformation
during these buckling tests. It also suggests that a large
area of the test specimen was susceptible to dimple formation when loaded close to the classical buckling pressure.

The high-speed motion picture frames of Fig.4.4 were taken
with a speed of 7000 frames/sec. The grid had 8 lines/cm.
The numbers mark the time in ms. The photographs confirm
the common opinion that, in complete spherical shells of
large r/t ratio, buckling begins with a rotationally symmetric inward dimple of a small central angle. The mandrel
does not have any effect on the initiation of the first
dimple, nor upon its growth until the bottom of the dimple
contacts the mandrel. At this moment, further inward motion
of the first dimple is restricted, and a large number of
small dimples develops rapidly at other sides. The formation
of these additonal dimples can be seen in the last several

210

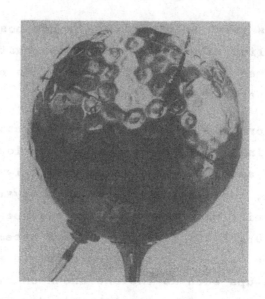

Fig.4.3   Nearly perfect sphere, buckled in the hard system
with mandrel.

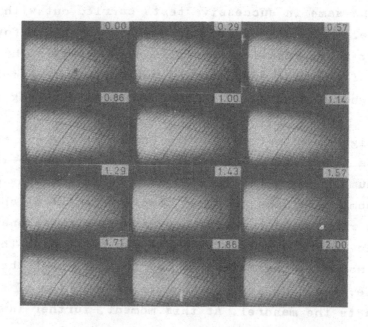

Fig.4.4   High-speed motion pictures of the buckling process
of a nearly perfect shell in the hard system with
mandrel (The numbers denote progress of time since
onset of buckling in ms).

frames of Fig.4.4 and much more clearly in the film taken
by Berke and Carlson [2].

The second type of tests in the hard system was conducted
with the wax mandrel removed from the test specimen. Due
to the control achieved in the hard system the deflections
could be kept small. Hence, the thin-walled nickel test
specimens did not undergo plastic deformations except occasional very localized ones; thus the test specimens could
be used for repeated tests. The primare purpose of these
tests was to obtain pressure-volume increment data in the
postbuckling range. Curves were obtained on the one hand
by loading into the postbuckling range followed by unloading
and on the other hand by manually inducing a dimple at a
sufficiently high value of the volume-increment to cause
the induced dimple to be stable. We spoke of a sufficently
high value of the volume-increment and not of a sufficeintly
high external pressure, because the set-up was volume controlled.

It is important to note that the postbuckling patterns always exhibit only one distinct buckle. The curves obtained
during unloading are shown in Fig.4.5. The black symbols
indicate the shape of the buckle. In the deep postbuckling
region the buckle has five corners. The number of corners
is reduced during unloading. Just before the buckle snaps
out to the unbuckled state its shape is circular. It is
remarkable that for each number of corners there is a corresponding separate postbuckling curve. This is similar to
the behavior of the circular cylinder, which shows separate
postbuckling curves for different circumferential wave numbers.

The pressure level of the postbuckling equilibrium states is
in the order of 10 % of the theoretical buckling pressure.

212

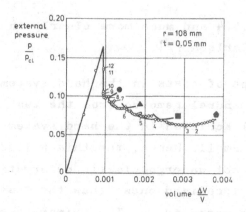

Fig. 4.5 Postbuckling curves obtained
in the hard system without
mandrel .

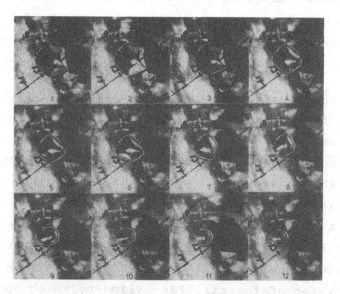

Fig.4.6 Stable postbuckling patterns obtained on a nearly
perfect sphere in the hard system without mandrel
(The numbers at the bottom of each frame correspond
to the numbers in the postbuckling diagram, Fig.4.5).

Fig.4.6 contains photographs of the stable postbuckling configurations as they appear during unloading. Each number below a buckle corresponds to a number in the postbuckling diagram, Fig.4.5; it correlates the mode shape with its location on the postbuckling curves.

High speed photographs were made also of the buckling of the specimens in the hard system with the wax mandrel removed, Fig.4.7. These photographs show that the buckle initiation is identical to the one shown on the first frames of Fig.4.4 for the system with mandrel. It is followed however by strong oscillations of the buckled area under the constraint of constant volume, and transitions between several mode shapes occur.

The photographs that were presented in Figs.4.6 and 4.7 have been shot during tests with a nearly perfect shell which buckled at a high pressure level. In Fig.4.8 you see dimples having appeared in a rather imperfect shell which buckled at a pressure level less than 50 % of the theoretical value. The dimples exhibit principally the same shape as found before with the nearly perfect sphere.

The experiments conducted in the soft loading system are performed under constant load. Both the value of the buckling pressure and the probable location of the initiation of buckling were known from previous tests in the hard system. Since no difference in the buckling pressure was expected between buckling in the hard and soft system, the spheres were loaded to within 10 - 15 % of their previous buckling pressures. The pressure was then raised by a few percent during 1 sec of the 2-sec running time of the camera which operated at a speed of about 2000 frames/sec. This procedure was repeated until buckling occurred during one of the load steps. It could be observed that there, again,

214

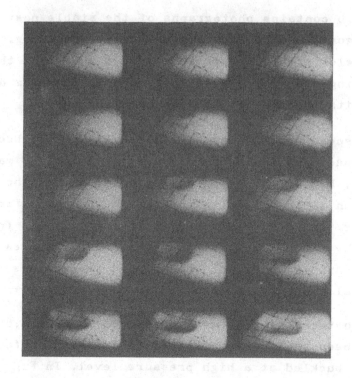

Fig.4.7   High speed motion pictures of the buckling process
          of a nearly perfect shell in the hard system with-
          out mandrel .

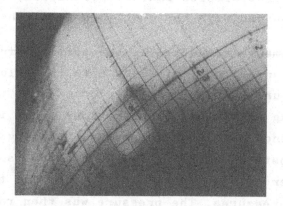

Fig.4.8a High speed motion picture of the buckling process
          of an imperfect shell in the hard system without
          mandrel .

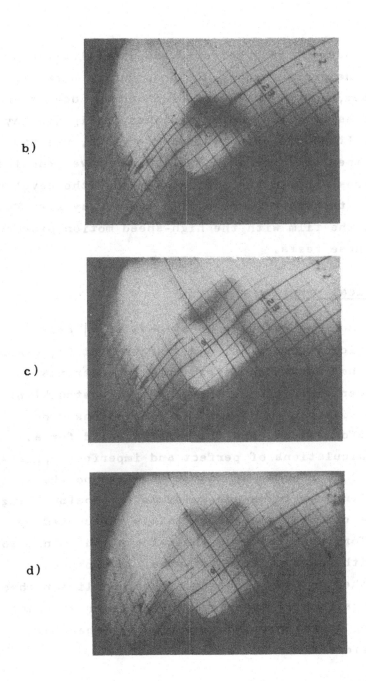

b)

c)

d)

Fig.4.8 High speed motion pictures of the buckling process
(continuation) of an imperfect shell in the hard
system without mandrel .

the first displacement mode detected was a small circular
dimple. Since the pressure does not drop during buckling
in a soft system, continued inward displacement occurs and
the various modes previously distinguished in Fig.4.6 may
be identified. It should be noted, however that, whereas
the maximum number of corners in Fig.4.6 was five, the last
several exposures shot in the soft system show the development of modes with higher numbers of corners, say six. This
can be seen in the film with the high-speed motion pictures
taken during these tests.

4.2.4 Theoretical results

The classical buckling calculation by Zoelly [4] starts
from axisymmetric deformations, and van der Neut [5] demonstrates later that non-axisymmetric buckling patterns do
not provide lower eigenvalues. The classical postbuckling
calculation by v.Kármán and Tsien [6] is also based on
axisymmetric deformations, and the same is valid for all
postbuckling calculations of perfect and imperfect spherical shells [7,8], performed up to the advanced postbuckling
region. - The tendency to assume axisymmetric postbuckling
patterns in the calculations was certainly influenced by
the fact that for these patterns the calculation is not so
laborious as with non-symmetric patterns. In a paper by
Hoff [9] published in 1969 a survey of the non-linear theoretical investigations of spherical shells under external
pressure loading is given without mentioning other than
axisymmetric deformations.

The fact that the perfect spherical shell would buckle in
a non-axisymmetric way was discovered only when the initial
postbuckling region was incorporated in the investigations.
Koiter [10] and Hutchinson [11] found that combinations of

axisymmetric and non-axisymmetric modes yield higher instability both at the bifurcation point itself and in the
initial postbuckling stage. Unfortunately neither Koiter
nor Hutchinson give a graphic representation of the theoretical non-symmetric buckling and postbuckling patterns.

Koiter [10] works with a series of spherical surface har-
monics, and it can not easily be understood which geometric
reality is hidden behind the short hand notation. At one
passage he mentions that his and Hutchinson's solutions exhibit hexagonal symmetry and at another passage he writes:
"The striking agreement between the deflection patterns,
observed in tests in which the magnitude of inward deflections was restricted by a mandrel with a very narrow gap
(cf. Fig.4.2), and the deflection patterns obtained theoretically suggests that we have, indeed, arrived at a proper
understanding of non-symmetric buckling".

It seems to us, that the mooncrater landscape which resulted
from the buckling test with mandrel does not appear to be
a convincing proof for non-symmetric buckling modes in the
initial postbuckling stage, because the sphere in this situation has no possibility to form one rather deep axisymmetric
buckle.

Hutchinson [11] confines his considerations upon a shallow
part of the spherical surface. In calculating buckling loads
he works with a chess-board pattern

$$w = \cos\left(k_x \cdot \frac{x}{r}\right) \cdot \cos\left(k_y \cdot \frac{y}{r}\right)$$

and found that the classical buckling pressure is associated
with any combination of wave numbers k_x and k_y satisfying the condition

$$k_x^2 + k_y^2 = \sqrt{12\,(1-\nu^2)}\,\frac{r}{t}\;.$$

In performing the postbuckling calculations he couples
eigenmodes corresponding to the condition shown above. This
study does not relate theoretical results to experimental
findings.

Postbuckling calculations the results of which can immediately be compared with test results so far were performed
only for axisymmetric patterns. In the following a paper
by Thompson [7] published in 1962 will be discussed. The
author made experimental and theoretical investigations
on two polyvinyl chloride shells; radius r = 53.3 mm ,
wall-thickness t = 2.54 mm , Young's modulus E = 316 N/cm^2,
Poisson's ratio ν = 0.48 .

The analysis used in these calculations is applicable with
good accuracy only to thin-walled spherical shells. The
sphere treated by Thompson is by no means thin-walled,
since its radius to wall-thickness ratio r/t = 21 . Thus
the following correlations must be viewed with some reser-
vation.

A theoretical and an experimental postbuckling curve are
plotted in Fig.4.9, and it can be seen that the agreement
is fairly good. The calculation was based on a perfect
shell. The Ritz method was employed using a polynominyl
expression with four free parameters as assumption for the
postbuckling pattern.

In Fig.4.10 theoretical and experimental postbuckling curves
for an imperfect sphere are compared with each other. For
the decay of the initial imperfection, a reasonable model

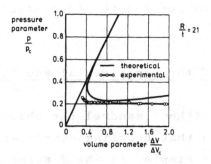

Fig.4.9   Load-volume curves of a nearly perfect sphere .

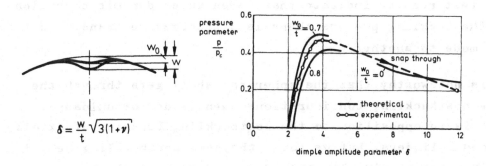

Fig.4.10   Load-deformation curves of an imperfect sphere .

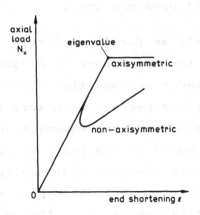

Fig.4.11   Different load-shortening curves of an axially
loaded cylinder.

was set up, the adequate choice of which surely contributed
to the good agreement between the theoretical and experimental curve.

4.2.5 Discussion of the experimental and theoretical results

In the hard system without mandrel the observations made during loading indicate that buckling starts with the formation
of a circular inward dimple. As the displaced volume increases and the shape changes; the dimple becomes cornered.
During unloading the transition from one shape to another is
the same, but in the opposite direction. In the soft system
the test results indicate that, even under dynamic conditions
of the buckling process there is a progressive transition from
one mode to another.

It is noteworthy that the spherical shell gets through the
same postbuckling configurations when loaded or unloaded.
This is in opposition to the postbuckling behavior of axially
loaded cylinders. But we found the same postbuckling behavior
in the case of circular cylinders under external pressure.
This behavior seems to be characteristic of all thin-walled
shells under external pressure load.

For the sphere as well as for the cylinder, each buckling
process begins with the appearance of a single buckle. In
the case of the cylinder we know that the buckling and postbuckling patterns of the perfect structure are periodic
around the circumference and consequently that the single
buckle must be attributed to the presence of initial imperfections. In a slightly advanced postbuckling stage the
influence of the initial imperfection gets weaker, and the
theoretical postbuckling patterns of the perfect cylinder
become dominating.

The complete thin-walled sphere behaves in an analoguous
manner. From the work of Koiter [10] and Hutchinson [11]
we know that the theoretical buckling pattern of the perfect sphere is non-axisymmetric. Hence, the axisymmetric
single buckle with which the buckling process of the real
sphere begins, as revealed by the high-speed motion pictures, must be attributed to an initial imperfection. In
the more advanced postbuckling state the influence of this
imperfection gets weaker, and the non-axisymmetric postbuckling pattern theoretically predicted for the initial
postbuckling region of the perfect shell can be observed.

But the analogy in the buckling and postbuckling behavior of
cylinders and spheres is not quite complete, since the imperfect sphere buckles with an axisymmetric pattern, whereas the cylinder always buckles in a non-axisymmetric mode.
This different behavior can easily be explained by the
knowledge of the theoretical postbuckling curves of the
perfect structures. Truly, in the cylinder as in the sphere,
there are axisymmetric and non-axisymmetric eigenmodes,
which for each structure belong to the same eigenvalue. At
the cylinder the load shortening curve does not drop in the
postbuckling region for the axisymmetric pattern, but drops
strongly for the non-axisymmetric mode, Fig.4.11. The sphere
however has postbuckling loads far below the buckling load
for axisymmetric as well as for non-axisymmetric patterns.

From the fact that the buckling pattern of the perfect shell
is not identical with the buckling pattern of the imperfect
structure it follows that the slope of the postbuckling curve
of the perfect sphere in the initial postbuckling region cannot play a vital role in the triggering of the buckling phenomenon of imperfect spheres. This conclusion might provide

the key for explaining, why spherical shells have so high
buckling loads, in spite of their low postbuckling loads
in the advanced postbuckling region.

It would be a fascinating and probably solvable task to
calculate postbuckling curves from the bifurcation point
into the advanced postbuckling region for axisymmetric as
well as for non-axisymmetric patterns, to compare them
with test results and with each other and thus to come to
a clear understanding of the buckling and postbuckling behavior of the pressure-loaded sphere. For this extensive
nonlinear shell calculation it will be of great help, that
the postbuckling patterns to be expected are known at least
for the initial postbuckling region from Koiter's asymptotic
analysis.

4.3 Spherical caps

4.3.1 Buckling investigations

The spherical cap is a doubly curved shell with edges at
which boundry conditions have to be taken into account and
at which boundary disturbances can occur. Hence, all dis-
agreeable phenomena we met with the sphere and the cylinder
will be found in the spherical cap. In the case of sufficiently long cylinders it was admissible to neglect the
influence of the edges on the state of stress and deformation in the prebuckling region and to base the calculations
on a pure membrane state of stress. In the case of the spherical cap, this simplification would lead to a wrong theoretical buckling load, since the axisymmetric deformations
occurring before buckling are of finite magnitude.

If the prebuckling deformations and the corresponding stresses

are taken into account exactly in the buckling analysis,
theoretical buckling loads are obtained [12,13], which are
in good agreement with the experimental buckling loads.
Weinitschke [13] in his paper, published in 1965, gives
a survey on his calculation method and the theoretical and
experimental results, published till this date. Fig.4.12
is quoted from this paper. The buckling pattern depends on
the parameter

$$\mu = \frac{b}{r} \cdot \sqrt{\frac{R}{t}} \cdot \sqrt{12\,(1-\nu^2)} \ .$$

If the value of this parameter is small, i.e. if the spherical cap is shallow or thick-walled, it buckles axisymmetri-
cally and the buckling load depends on the opening angle.
If $\mu > 5$, the buckling pattern is non-symmetric, and the
buckling load is almost independent of μ . From Fig.4.12
one can see that the calculated and the measured values
agree well.

4.3.2 Postbuckling investigations

Klöppel and Jungbluth [17] published a sensational paper in
1953 in which the buckling and postbuckling behavior of
spherical caps with clamped edges was investigated experimentally. The test specimens were manufactured by deep
forming out of a steel sheet. They have the following dimensions: radius of sphere R = 520 mm or 250 mm , radius
of spherical cap always r = 200 mm , wall-thickness between t = 0.3 mm and 0.6 mm .

Klöppel's snap-through loads are considerably lower than
the theoretical buckling loads; the comparison is shown in
Fig.4.13. This discrepancy may be due to the fact that the
die for the deep forming of the test specimens was not in

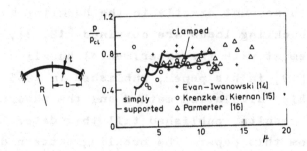

Fig.4.12   Theoretical and experimental buckling loads
spherical cap [13].

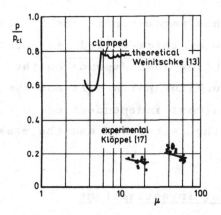

Fig.4.13   Comparison of Klöppel's experimental and
Weinitschke's theoretical results.

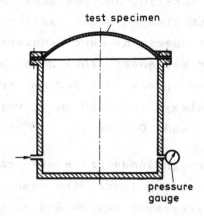

Fig.4.14   Klöppel's test device.

the correct shape. Fig.4.14 is a schematical representation
of the test facility with built-in test specimen. The edges
of the spherical cap are bent over so that they can be
clamped between the flanges of the test device. This edge
clamping is probably not imperfection free, and that may
be another reason for Klöppel's snap-through loads being
lower than the newer test results, presented in Fig.4.12.

Klöppel's paper has become famous mainly because, for the
first time, high-speed motion pictures of the buckling process of a thin-walled shell were published. Fig.4.15 shows
as an example the snap-through of a cap with sphere radius
R = 250 mm and wall-thickness t = 0.34 mm . The buckling
process begins with the jumping-in of a small buckle at the
edge. The buckle becomes larger, extends in an intermediate
state beyond the middle of the spherical cap and finally
grows as large as the whole spherical cap, that means, the
cap completely snaps through. The snap-through process lasts
10 ms .

The perfect shell buckles with several dimples uniformly
distributed over the periphery. This theoretical statement
is confirmed by the test result in so far as no axisymmetric
postbuckling pattern appears except at the very end of the
snap-through. The postbuckling patterns appearing in the
very initial stages are not entirely revealed by the high-
speed motion pictures. It can not be excluded that an unstable postbuckling pattern consisting of several buckles
has existed, but only for a short time or with amplitudes
so small that it could not be detected by the high-speed
camera. With some good will one can see several edge buckles
in the first frame of Fig.4.15.

Fig.4.16 shows an experimental load deformation curve of the

226

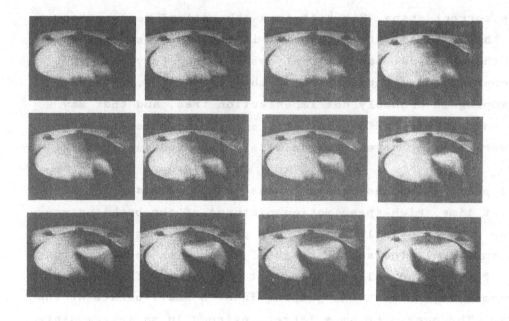

Fig.4.15   High speed motion pictures of the initial
postbuckling stage.

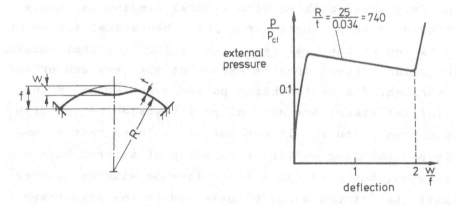

Fig.4.16   Load deformation curve of the spherical
cap   R = 250 mm ,   t = 0.34 mm.

buckling and postbuckling process. The abscissa represents the ratio

$$\frac{w}{f} = \frac{\text{vertical deflection of apex}}{\text{height of the cap}} .$$

When the spherical cap has snapped-through completely, a new stable equilibrium state is attained, in which the spherical cap is subjected to tensile stresses. Now the load can be increased further till the ultimate stress limit is reached.

The load drop during the snap-through, seen in Fig.4.16, has nothing to do with the load carrying capacity of the shell. It occurs, because the pressure difference loading the shell is reduced by a decrease of the evacuated volume in the buckling device during the fall-in of the buckles.

Literature

[1]   CARLSON, R.L.   Experimental Studies of the Buckling
        SENDELBECK, R.L.   of Complete Spherical Shells.
        HOFF, N.J.   Experimental Mechanics Vol.7 (1967),
                            pp.281-288

[2]   BERKE, L.   Experimental Studies of the Postbuck-
        CARLSON, R.L.   ling Behavior of Complete Spherical
                            Shells.

                            Experimental Mechanics Vo.8 (1968)
                            pp.548-553

[3]   THOMPSON, J.M.T.   Making of Thin Metal Shells for Model
                            Stress Analysis.

                            J. of Mech.Engn.Sci. Vol.2 (1960)
                            p. 105

[4]   ZOELLY, R.   Über ein Knickungsproblem an der
                            Kugelschale.

                            Dissertation, Zürich (1915)

228

[5]   NEUT, A. van der   The Elastic Stability of the Spherical Shell.

Thesis Delft, H.J. Paris, Amsterdam (1932)

[6]   KARMAN, Th. von   The Buckling of Spherical Shells by TSIEN, H.S.   External Pressure.

J. Aeron.Sci. Vol.7 (1939), pp.43-50

[7]   THOMPSON, J.M.T.   The Elastic Instability of a Complete Spherical Shell.

Aeron.Quart. Vol.13 (1962),pp.189-201

[8]   KOGA, T.   The Axisymmetric Snap-Through Buckling of Thin-Walled Spherical Shells. HOFF, N.J.

Internat. J. of Solids and Structures, Vol.5 (1969), pp.679-697

[9]   HOFF, N.J.   Some Recent Studies of the Buckling of Thin Shells.

Aeron. J. of Roy.Aeron.Soc. Vol.73 (1969), pp.1057-1070

[10]   KOITER, W.T.   The Nonlinear Buckling Problem of a Complete Spherical Shell under Uniform External Pressure.

Kon.Ned.Ak.Wet., Amsterdam, Proceedings, Series B, 72 (1969), No.1 and 2, pp.40-123

[11]   HUTCHINSON, J.W.   Imperfection Sensitivity of External Pressurized Spherical Shells.

J. Appl.Mech. Vol.34 (1967), pp.49-55

[12]   HUANG, N.C.   Unsymmetrical Buckling of Thin Shallow Spherical Shells.

J. Appl.Mech. Vol.31 (1964),pp.447-457

[13]   WEINITSCHKE, H.J.   On Asymmetric Buckling of Shallow Spherical Shells.

J. Math.Phys. Vol.44 (1965), pp.141-163

[14] EVAN-   Deformations and Stability of Spheri-
       IWANOWSKI, R.M.   cal Shells under Action of Concen-
       LOO, T.C.   trated Loads and Uniform Pressure.

                            Techn. Report 834 (11), No.4,
                            Syracuse University, Research
                            Institute, June (1962)

[15] KRENZKE, M.A.   Elastic Stability of Near-Perfect
       KIERNAN, T.J.   Shallow Spherical Shells.

                            AIAA J. Vol.1 (1963), pp.2855-2857

[16] PARMERTER, R.R.   The Buckling of Clamped Shallow
                            Spherical Shells under Uniform
                            Pressure.

                            Ph.D. Thesis, California Inst. of
                            Technology, Nov. (1963)

[17] KLÖPPEL, K.   Beitrag zum Durchschlagproblem dünn-
       JUNGBLUTH, O.   wandiger Kugelschalen.

                            Der Stahlbau, Band 22 (1953),S.121-130

5. ANALYSIS

5.1 Cylindrical shells

5.1.1 Notations

A_{ij} coefficients of extensional compliance, defined in equ. (15)

D_{ij} coefficients of flexural rigidity, defined in equs. (9) and (10)

E   Young's modulus

G   modulus of rigidity

M_x, M_y, M_{xy} bending, torsional moments per unit length of median surface, defined in Fig.5.2

N_x, N_y, N_{xy} membrane direct and shear forces per unit length of median surface, defined in Fig.5.2

$\bar{N}_x$ non-dimensional axial force per unit length, positive in compression, defined on page 106

Q_x, Q_y transverse shear forces, defined in Fig.5.2

a_{jm} coefficients of series for w, defined in equs. (27) and (31)

a_o radial displacement, defined in equ. (28)

$b_{\tilde{j}\tilde{m}}$ coefficients of series for ϕ, defined in equs. (27) and (31)

j, m   integers

l   length of the cylinder

$\bar{l}$ length parameter, defined on page 89 and 106

n   circumferential wave number

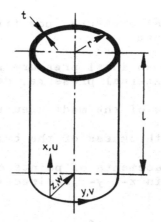

Fig.5.1 Co-ordinates and dimensions.

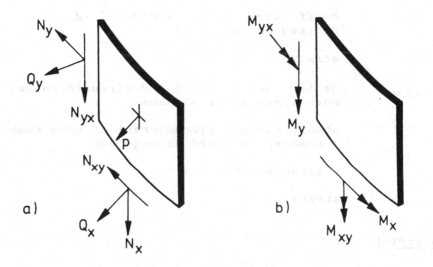

Fig.5.2 Stress resultants.

a. linear   b. nonlinear
Fig.5.3 Illustration of strain-displacement relations.

p	lateral pressure, positive as internal pressure
$\bar{p}$	nondimensional pressure parameter, positive for external pressure, defined on page 89
r	radius of the median surface of the cylinder
t	wall-thickness of the cylinder
u, v, w	displacements of points on the median surface in x-, y-, z-direction, defined in Fig.5.1
x, y, z	axial, circumferential and radial coordinates, defined in Fig.5.1
ϕ	stress function, defined in equ. (12)
$\alpha_{\tilde{j}\tilde{m}}$, $\beta_{\tilde{j}\tilde{m}}$	coefficients in the series of ϕ, defined in equ. (32)
ε	strain
ε_x, ε_y, γ_{xy}	median surface axial and circumferential normal and shear strains
η	nondimensional circumferential wave number parameter, defined on page 106
ν	Poisson's ratio
σ	stress

Superscripts

~	governs the stress function ϕ
−	governs the radial displacement w
*	denotes fixed indices in the Galerkin equ. (39)

5.1.2 Differential equations

The differential equations will be first derived for a thin-
walled circular cylinder and later simplified, so that they
become valid for plates. Fig.5.1 shows the cylinder and its
dimensions, as well as the coordinates and displacements
used for the following calculations. "Thin-walled" means
that the ratio

$$\frac{r}{t} > 100$$

is met. This is no exact limit, but for practical purposes
it is a rather useful assumption. The bigger the ratio   r/t
the better the validity of the shell theory, presented in
this paper and generally used in postbuckling calculations.

It is assumed that all deformations take place in the elastic range so that Hook's law

$$\sigma = E \cdot \varepsilon$$

is valid. This again is no strict condition, but a simplification reasonable for practical purposes.

In Fig.5.2 you see the forces and moments acting upon an
element of the shell. The stress resultants on the positive
edge of the element point in the positive directions. The
bending moments are positive for positive curvature, that
means for tension in the outside fibres of the shell. The
torsional moments are positive, if on the rectangular shell
element, Fig.5.2b, the corner with positive coordinates is
displaced into the positive direction, i.e. to the inside.
Internal pressure is positive.

It shall be assumed, as it is customary in the theory of
shells, that the shear deformations transverse to the

surface of the shell, i.e. the deformations caused by the transverse shear forces Q_x and Q_y , are negligibly small. Moreover the direct stresses and strains perpendicular to the surface of the shell will be neglected. From this it follows that the stresses are distributed linearly across the shell thickness. Within the frame of the simplifications specified above there are a great many shell theories which differ relative to the neglection or inclusion of terms of minor importance.

Postbuckling calculations require a nonlinear theory of shells since finite deformations have to be included. To keep the effort within reasonable limits one uses for these calculations the simplest of all shell theories offered in the literature, the so called "shallow shell theory". This theory was introduced in 1934 by Donnell [1].

Some years ago W. Flügge explained to the senior author the generation of Donnell's shallow shell theory on the following not quite serious manner: Donnell laid the available shell theories one upon another and held them against the light. Then he collected and used for his further calculations the terms which appeared transparent because they were contained in all theories.

In the mean time Donnel's theory has been thoroughly analysed in the literature and it has been found that it may be used under the following restrictions:

- The circumferential angle between two nodal points of the buckling or postbuckling pattern is assumed to be small. Hence, Donnell's theory yields useful results for circumferential wave numbers $n \leq 4$ only. This is no strict law, but an arbitrary somewhat vague

limitation. The bigger the number of waves on the
circumference the more exact the calculation.

The deformations u and v in longitudinal and
circumferential direction are small compared with
the radial deformation w . Due to this simplification, Donnell's theory fails totally if the cylinder
is of such a length that it buckles as a column,
since in this case the tangential displacement v
and the radial displacement w are equal.

On Fig.5.2 one can read immediately the conditions for equilibrium at the   u n d e f o r m e d   shell element [2].

Forces in x-direction

(1)
$$\frac{\partial N_x}{\partial x} + \frac{\partial N_{xy}}{\partial y} = 0 .$$

Forces in y-direction

(2)
$$\frac{\partial N_y}{\partial y} + \frac{\partial N_{xy}}{\partial x} - \frac{Q_y}{r} = 0 .$$

Forces in z-direction

(3)
$$\frac{\partial Q_y}{\partial y} + \frac{\partial Q_x}{\partial x} + \frac{N_y}{r} - p = 0 .$$

Moments about the x-axis

(4)
$$\frac{\partial M_y}{\partial y} + \frac{\partial M_{xy}}{\partial x} + Q_y = 0 .$$

Moments about the y-axis

(5)
$$\frac{\partial M_x}{\partial x} + \frac{\partial M_{xy}}{\partial y} + Q_x = 0 .$$

Moments about the z-axis

(6)
$$N_{xy} - N_{yx} - \frac{M_{xy}}{r} = 0 .$$

The crossed out terms are neglected in Donnell's theory.

The equilibrium conditions for the d e f o r m e d shell element contain the same terms and moreover, additional ones, which are products of the membrane forces and the deformations. Within the frame of Donnell's shell theory such additional terms are introduced in the equilibrium conditions for the forces in z-direction only. One obtains

from N_x and the curvature $\dfrac{\partial^2 w}{\partial x^2}$ the additional component

$$N_x \cdot \frac{\partial^2 w}{\partial x^2} \quad ;$$

from N_y and the curvature $\dfrac{\partial^2 w}{\partial y^2}$ the additional component

$$N_y \cdot \frac{\partial^2 w}{\partial y^2} \quad ;$$

and from N_{xy} and the twist $\dfrac{\partial^2 w}{\partial x \cdot \partial y}$ the additional component

$$2 N_{xy} \cdot \frac{\partial^2 w}{\partial x \cdot \partial y} \quad .$$

With these supplements the equilibrium condition of the
forces in z-direction becomes

(7)
$$\frac{\partial Q_y}{\partial y} + \frac{\partial Q_x}{\partial x} + \frac{N_y}{r} +$$
$$+ N_x \cdot \frac{\partial^2 w}{\partial x^2} + N_y \cdot \frac{\partial^2 w}{\partial y^2} + 2 N_{xy} \cdot \frac{\partial^2 w}{\partial x \partial y} - p = 0.$$

If in this equation the shear forces Q_x and Q_y are
eliminated by using equations (4) and (5), then follows

(8)
$$\frac{\partial^2 M_x}{\partial x^2} + 2 \frac{\partial^2 M_{xy}}{\partial x \partial y} + \frac{\partial^2 M_y}{\partial y^2} - \frac{N_y}{r} -$$
$$- N_x \cdot \frac{\partial^2 w}{\partial x^2} - N_y \cdot \frac{\partial^2 w}{\partial y^2} - 2 N_{xy} \cdot \frac{\partial^2 w}{\partial x \partial y} + p = 0.$$

Within the frame of Donnell's theory one can express the
moments by the radial displacement w [3]

(9a-c)
$$M_x = D_{11} \cdot \frac{\partial^2 w}{\partial x^2} + D_{12} \cdot \frac{\partial^2 w}{\partial y^2} \quad ,$$
$$M_y = D_{12} \cdot \frac{\partial^2 w}{\partial x^2} + D_{22} \cdot \frac{\partial^2 w}{\partial y^2} \quad ,$$
$$M_{xy} = \qquad D_{33} \cdot 2 \frac{\partial^2 w}{\partial x \partial y} \quad .$$

For isotropic cylinders the coefficients of flexural rigidity are

(10)
$$D_{11} = D_{22} = \frac{E \cdot t^3}{12 (1 - \nu^2)} \quad ,$$
$$D_{12} = \nu \cdot D_{11} = \nu \cdot \frac{E \cdot t^3}{12 (1 - \nu^2)} \quad ,$$
$$D_{33} = \frac{E \cdot t^3}{24 (1 + \nu)} .$$

Introduction of equation (9) in equation (8) yields

$$D_{11} \cdot \frac{\partial^4 w}{\partial x^4} + 2\left(D_{12}+D_{33}\right)\frac{\partial^4 w}{\partial x^2 \partial y^2} + D_{22}\frac{\partial^4 w}{\partial y^4} - \frac{N_y}{r} + p$$

(11)
$$- N_x \cdot \frac{\partial^2 w}{\partial x^2} - N_y \cdot \frac{\partial^2 w}{\partial y^2} - 2N_{xy} \cdot \frac{\partial^2 w}{\partial x \partial y} = 0 \ .$$

By introducing a stress function ϕ defined as follows

(12) $\qquad \dfrac{\partial^2 \Phi}{\partial y^2} = N_x \qquad \dfrac{\partial^2 \Phi}{\partial x^2} = N_y \qquad -\dfrac{\partial^2 \Phi}{\partial x \partial y} = N_{xy} = N_{yx}$

the equilibrium conditions (1), (2) and (6) are satisfied.

If in equation (11) the membrane forces are substituted by the stress function ϕ , defined in (12), the condition for equilibrium in the z-direction is obtained as

(13)
$$D_{11} \cdot \frac{\partial^4 w}{\partial x^4} + 2\left(D_{12}+2D_{33}\right)\frac{\partial^4 w}{\partial x^2 \partial y^2} + D_{22} \cdot \frac{\partial^4 w}{\partial y^4} - \frac{1}{r} \cdot \frac{\partial^2 \Phi}{\partial x^2} + p -$$

$$- \frac{\partial^2 \Phi}{\partial y^2} \cdot \frac{\partial^2 w}{\partial x^2} - \frac{\partial^2 \Phi}{\partial x^2} \cdot \frac{\partial^2 w}{\partial y^2} + 2\frac{\partial^2 \Phi}{\partial x \partial y} \cdot \frac{\partial^2 w}{\partial x \partial y} = 0 \ .$$

This is a non-linear partial differential equation for the unknown functions w (x,y) and ϕ (x,y) . To determine these functions we still need a second equation. For this we cannot apply the equilibrium conditions (1-6) since we used them already at the introduction of the stress function (12) and at establishing the equilibrium condition (13).

The second equation for w and ϕ is obtained by using

the condition for compatibility, i.e. the condition that
the deformations caused by the membrane forces have to be
compatible. The strains ε_x, ε_y, γ_{xy} can be expressed
on the one hand, by the membrane forces N_x, N_y and N_{xy}
and, on the other hand, by the displacements u, v, w.

The deformations due to the membrane forces are compatible
if the term

(14)
$$\frac{\partial^2 \varepsilon_x}{\partial x^2} + \frac{\partial^2 \varepsilon_y}{\partial y^2} - \frac{\partial^2 \gamma_{xy}}{\partial x \partial y}$$

first expressed by the membrane forces, i.e. by the stress-
function ϕ, and second expressed by the displacement u,
v, w, is of the same value. Since in this manipulation
the terms with u and v are cancelled, the condition of
compatibility will be an equation containing only the two
unknowns w and ϕ.

The connection between the strains and the membrane forces
is expressed in the law of elasticity [3]

$$\varepsilon_x = A_{11} \cdot N_x + A_{12} \cdot N_y \ ,$$

(15a-c)
$$\varepsilon_y = A_{12} \cdot N_x + A_{22} \cdot N_y \ ,$$

$$\gamma_{xy} = \qquad\qquad A_{33} \cdot N_{xy} \ .$$

For isotropic cylinders the coefficients of extensional
compliance are

$$A_{11} = A_{22} = \frac{1}{E \cdot t} \ ,$$

(16)
$$A_{12} = -\nu \cdot A_{11} = -\frac{\nu}{E \cdot t} \ ,$$

$$A_{33} \qquad\qquad = \frac{2\,(1+\nu)}{E \cdot t} \ .$$

240

The elasticity law (15a-c) is independent of whether the deflections are finite or infinite. But in the geometric relations additional nonlinear terms have to be considered in the case of finite deflections

$$\varepsilon_x = \frac{\partial u}{\partial x} + \frac{1}{2}\left(\frac{\partial w}{\partial x}\right)^2$$

(17a-c)

$$\varepsilon_y = \frac{\partial v}{\partial y} - \frac{w}{r} + \frac{1}{2}\left(\frac{\partial w}{\partial y}\right)^2$$

$$\gamma_{xy} = \frac{\partial u}{\partial y} + \frac{\partial v}{\partial x} + \frac{\partial w}{\partial x}\cdot\frac{\partial w}{\partial y}\ .$$

In the equation (17a) for ε_x the first term of the right-hand side contains the change in the length of the shell element in the direction of the undeformed x-axis, Fig.5.3a; if the shell element dx is elongated by $(\partial u /\partial x)\cdot dx$ it undergoes a strain $\partial u /\partial x$. The second term is nonlinear and will be considered only if finite deflections are to be included into the calculation. It contains the influence of the rotation upon the strain. On Fig.5.3b one can immediately read that

$$ds = \sqrt{dx^2 + \left(\frac{\partial w}{\partial x}\cdot dx\right)^2} = \sqrt{1 + \left(\frac{\partial w}{\partial x}\right)^2}\ dx\ .$$

Hence, for the part of the strain which results from the rotation follows

$$\Delta\varepsilon_x = \frac{ds - dx}{dx} = \sqrt{1 - \left(\frac{\partial w}{\partial x}\right)^2} - 1 = \sim \frac{1}{2}\left(\frac{\partial w}{\partial x}\right)^2 .$$

In the expression for ε_y you see the corresponding terms as in the expressiong for ε_x , and in addition one more term w/r which expresses by how much the circumference

of the cylinder is being compressed if the wall is displaced by the distance w radially toward the inside.

Also in the expression for the shear strain γ_{xy} there are
linear terms which have to be considered in all cases and
one nonlinear term which is to be considered in the case
of finite deflections only.

If on the one hand the elasticity law (15) with the stress
function (12), and on the other hand the geometric relations (17) are inserted into the combination of strain
derivatives (14), and the two expressions thus obtained
are set equal to each other the condition for the compatibility is obtained as:

(18)
$$A_{11} \cdot \frac{\partial^4 \Phi}{\partial y^4} + \left(2A_{12}+A_{33}\right) \frac{\partial^4 \Phi}{\partial x^2 \partial y^2} + A_{22} \cdot \frac{\partial^4 \Phi}{\partial x^4}$$
$$= -\frac{1}{r}\frac{\partial^2 w}{\partial x^2} + \left(\frac{\partial^2 w}{\partial x \partial y}\right)^2 - \frac{\partial^2 w}{\partial x^2} \cdot \frac{\partial^2 w}{\partial y^2} .$$

This also is a nonlinear partial differential equation for
the two unknowns w and ϕ . The two differential equations
(13) and (18) and the boundary conditions describe the problem. Since the differential equations are nonlinear,there
are infinitely many solutions. The question which ones of
these are physically reasonable will be answered in chapter 5.8 .

If the stiffness of the shell is constant over the entire
surface the coefficients A_{ij} and D_{ij} are constant.
Hence, the differential equations (13) and (18) have constant coefficients which is important for the methods of
solution.

For axisymmetric equilibrium states equation (13) is simplified to

(19)
$$D_{11} \cdot \frac{\partial^4 w}{\partial x^4} - \frac{1}{r} \cdot N_y + p - N_x \cdot \frac{\partial^2 w}{\partial x^2} = 0 .$$

If loads in axial direction act at the edges only N_x is
constant along the whole cylinder, whereas w and N_y may
be variable in the direction of the axis.

The second condition for calculating w and N_y is obtained from the equations (15b) and (17b) as

(20)
$$A_{12} \cdot N_x + A_{22} \cdot N_y + \frac{w}{r} = 0 .$$

For the calculation of plates, i.e. for $r = \infty$ the equations (13) and (18) are simplified to

(21)
$$D_{11} \cdot \frac{\partial^4 w}{\partial x^4} + 2 \left(D_{12} + 2 D_{33} \right) \frac{\partial^4 w}{\partial x^2 \partial y^2} + D_{22} \cdot \frac{\partial^4 w}{\partial y^4} + p -$$
$$- \frac{\partial^2 \Phi}{\partial y^2} \cdot \frac{\partial^2 w}{\partial x^2} - \frac{\partial^2 \Phi}{\partial x^2} \cdot \frac{\partial^2 w}{\partial y^2} + 2 \cdot \frac{\partial^2 \Phi}{\partial x \partial y} \cdot \frac{\partial^2 w}{\partial x \partial y} = 0$$

(22)
$$A_{11} \cdot \frac{\partial^4 \Phi}{\partial y^4} + \left(2 A_{12} + A_{33} \right) \frac{\partial^4 \Phi}{\partial x^2 \partial y^2} + A_{22} \cdot \frac{\partial^4 \Phi}{\partial x^4}$$
$$= \left(\frac{\partial^2 w}{\partial x \partial y} \right)^2 - \frac{\partial^2 w}{\partial x^2} \cdot \frac{\partial^2 w}{\partial y^2} .$$

In 1939 Marguerre [6] derived the equations of the shallow
shell theory as Eulerian equations from the principle of

stationary potential energy. His results contained Donnell's equations as a special case and could thus serve to verify Donnell's theory. It may be mentioned that Marguerre's basic geometric relations appear in a somewhat different form, because he treated the shallow shell as a curved plate.

As Koiter [7] has shown, the equations of the shallow shell theory under certain conditions are even valid for non-shallow shells. The function w (x,y) - which in the shallow shell theory is considered as the normal deflection - has then to be interpreted as a "curvature function". Koiter proposed to speak of a theory for "quasi-shallow shells", or of a theory for "shells of small Gaussian curvature".

An extension of equations (13) and (18) for orthotropic, eccentrically stiffened cylinders can be found in [5]. In this paper two variants of the Donnell differential equations are presented. In the first variant the equilibrium state is described by the radial displacement w and the stress-function ϕ as is done in this paper; in the second variant it is represented by the three displacements u , v , w . The first variant is better suited for postbuckling calculations than the second.

The derivation of the equations is again based on the geometric simplifications of the shallow shell theory. The potential energy of the shell is first formulated in terms of displacements, and the equilibrium equations are established as Eulerian differential equations following from the variational principle of the potential energy having a stationary value at an equilibrium state.

5.1.3 Survey on some proven solution methods

The differential equations (19) and (20) for axisymmetric

244

deformations are exactly solved by means of the function

(23)
$$w = \sum_{i=1}^{i=4} a_i \cdot e^{\lambda_i \frac{x}{r}}$$

$$\Phi = \sum_{i=1}^{i=4} b_i \cdot e^{\lambda_i \frac{x}{r}} \ .$$

The decay factors λ_i result from the characteristic equation. The integration constants are defined by the boundary conditions.

If the equilibrium state is not axisymmetric, no exact solutions are known for the differential equations (13) and (18). It is usual to write down series expansions for the solutions the individual terms of which satisfy the boundary conditions

(24)
$$w = \sum \sum a_{jm} \cdot f_m(x) \cdot f_j(y)$$

$$\Phi = \sum \sum b_{\tilde{j}\tilde{m}} \cdot f_{\tilde{m}}(x) \cdot f_{\tilde{j}}(y) \ ,$$

with the coefficients a_{jm} and $b_{\tilde{j}\tilde{m}}$ to be determined so that the differential equations (13) and (18) are satisfied as well as possible. By the choice of the series terms one can in general determine the resulting postbuckling pattern. The postbuckling pattern presented in Fig.5.4b is obtained using a series with $n = 9$ waves in circumferential direction and $m = 2$ half waves in longitudinal direction, and with the test boundary conditions

(25) $\quad \dfrac{\partial u}{\partial y} = 0 \qquad\qquad v = 0 \qquad\qquad w = 0 \qquad\qquad \dfrac{\partial w}{\partial x} = 0 \ .$

245

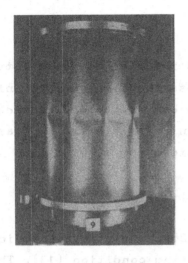

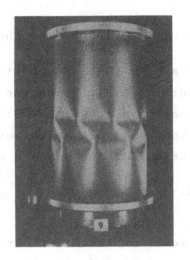

a. one-tier   b. two-tier

Fig.5.4   Postbuckling patterns of an axially loaded
          cylinder.

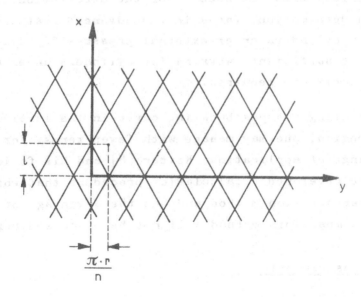

Fig.5.5   Periodicity of diamond pattern.

The calculation is carried through in three steps.

- The series (24) is substituted into the compatibility condition (18). Since the stress function appears only linearly in equation (18), one can solve the equation so that one obtains the unknown coefficients $b_{\tilde{j}\tilde{m}}$ as functions of the unknown coefficients a_{jm}

(26)
$$b_{\tilde{j}\tilde{m}} = F\left(a_{jm}\right).$$

- By means of the relationship (26), the stress function ϕ is eliminated from the equilibrium condition (13). This manipulation produces a nonlinear equation system for the unknowns a_{jm}.

- The nonlinear equation system is solved by iteration.

30 - 50 series terms are needed for the description of stress and deformation states in the advanced postbuckling region. For cylinders under external pressure 30 series terms may be sufficient, whereas for cylinders under axial load more terms are required.

When calculating the postbuckling curve in the initial post-buckling region, one may manage with fewer terms. For this limited range of application, Koiter [8] and his followers [9,10] have developed a calculation procedure that works with the perturbation method and has the advantage of short computing times. This method will not be discussed here.

5.1.4 Series expansion

For the diamond-shaped postbuckling pattern that extends endlessly over an i n f i n i t e l y l o n g cylin-der, the following series is used

$$w = \sum \sum a_{jm} \cdot \cos \frac{m\pi x}{l} \cdot \cos \frac{jny}{r} + a_0$$

(27a-b)

$$\Phi = \sum \sum b_{\tilde{j}\tilde{m}} \cdot \cos \frac{\tilde{m}\pi x}{l} \cdot \cos \frac{\tilde{j}ny}{r} + \overset{\circ}{N}_x \frac{y^2}{2} + p \cdot r \frac{x^2}{2} ,$$

with $\quad$ j = 0, 1, 2, 3 $\quad$ and $\quad \tilde{j}$ = 0, 1, 2, 3

$\qquad$ m = 0, 1, 2, 3 $\quad$ and $\quad \tilde{m}$ = 0, 1, 2, 3

$\qquad$ l = axial half wave length

$\qquad \overset{\circ}{N}_x$ = uniformly distributed axial force

$\qquad$ pr = uniformly distributed circumferential force

$\qquad a_0$ = uniformly distributed radial displacement

(28)
$$a_0 = - \left(A_{12} \overset{\circ}{N}_x + A_{22} \, p \cdot r \right) \cdot r .$$

The circumferential wave number $\quad$ n $\quad$ can be prescribed to get the desired postbuckling curve.

For the axial half wave number $\quad$ m $\quad$ a more complicated pro-cedure is required. Fig.5.5 shows that the buckles are so staggered that after half a buckle length in longitudinal and circumferential direction, the same radial displacement occurs again.

(29)
$$w \left(x = 0, \, y = 0 \right) = w \left(x = l, \, y = \frac{\pi r}{n} \right) .$$

Applying this condition to the series (27a) yields, for each term, the relation

$$\cos 0 \cdot \cos 0 \; = \cos m\pi \cdot \cos j\pi ,$$

which is fullfilled, if

(30)   j + m =   even .

Only the terms of the series (27) that satisfy the condition (30) are introduced in the calculation.

For cylinders of finite length the calculations are much
more complicated. They can be simplified considerably, if
the test boundary conditions (25) are not strictly satisfied, but modified by allowing of the radial displacement
a_o which is defined in equation (28). The following series
has proved useful:

$$w = \sum \sum a_{jm} \cdot \left(\cos \frac{m \pi x}{l} - \cos \frac{(m+2) \pi x}{l} \right) \cos \frac{jny}{r} + a_o$$

$$(31) \quad \Phi = \sum \sum b_{\tilde{j}\tilde{m}} \left(\alpha_{\tilde{j}\tilde{m}} \cdot \cos \frac{\tilde{m} \pi x}{l} - \beta_{\tilde{j}\tilde{m}} \cdot \cos \frac{(\tilde{m}+2) \pi x}{l} \right) \cdot \cos \frac{\tilde{j}ny}{r}$$

$$+ \overset{\circ}{N}_x \frac{y^2}{2} + pr \cdot \frac{x^2}{2} \; ,$$

with   j = 0, 1, 2, 3 ...   and $\tilde{j}$ = 0, 1, 2, 3 ...

          l = length of cylinder

$\overset{\circ}{N}_x$, pr , a_o as in equation (28). $\alpha_{\tilde{j}\tilde{m}}$ and $\beta_{\tilde{j}\tilde{m}}$ are calculated by means of equations (15), (17) and (12) from the
boundary condition v = 0 .

$$\alpha_{\tilde{j}\tilde{m}} = \frac{A_{12}}{A_{22}} \cdot \left(\frac{\tilde{j}n}{r} \right)^2 + \left(\frac{(\tilde{m}+2)\pi}{l} \right)^2 ,$$

(32)

$$\beta_{\tilde{j}\tilde{m}} = \frac{A_{12}}{A_{22}} \cdot \left(\frac{\tilde{j}n}{r} \right)^2 + \left(\frac{\tilde{m}\pi}{l} \right)^2 .$$

The values m for the half wave number in longitudinal
directinn depend on whether a one-tier postbuckling pattern,
Fig.5.4a, or a two-tier postbuckling pattern, Fig.5.4b, is
to be calculated. For one-tier patterns m has the values
0, 2, 4, 6 whereas for two-tier postbuckling patterns m
follows from the condition (30).

By the foregoing exposition the scope of the series terms
is defined. Now the sequence and the number of the terms
included have to be determined. Fig.5.6 shows the scheme
used for the diamond-shaped v.Kármán pattern of the infinitely long cylinder. In this table the relevant terms,
i.e. the terms satisfying condition (30) are arranged. The
terms of the top left hand corner, with exception of the
term j = 0 , m = 0 , are included in the calculation in
such a sequence that the corner region gradually becomes
larger. The shaded area in Fig.5.6 indicates the 50 terms
generally used in our calculations. The sequence was
searched and found by trial and error, so that the calculation converged.

For isotropic axially compressed cylinders we always used
a 50-term series. Occasional checks with 60 terms or computations with double precision gave the same results, and
thus the reliability of our calculations was verified.

The choice of the series defines the postbuckling curve
which is to be calculated, but nothing has been said until
no about the point on the curve which is aimed at by the
iteration. It seems appropriate to fix this point by pre-
scribing the postbuckling load. But this procedure proved
a failure, because there can be three equilibrium states
for one load, Fig.5.7, and according to our experience, the
iteration would prefer the equilibrium state on the prebuckling line, which is not the one we are looking for.

j	m									
0	0	2	4	6	8	10	12	14	16	18
1	1	3	5	7	9	11	13	15	17	19
2	0	2	4	6	8	10	12	14	16	18
3	1	3	5	7	9	11	13	15	17	19
4	0	2	4	6	8	10	12	14	16	18
5	1	3	5	7	9	11	13	15	17	19
6	0	2	4	6	8	10	12	14	16	18
7	1	3	5	7	9	11	13	15	17	19
8	0	2	4	6	8	10	12	14	16	18
9	1	3	5	7	9	11	13	15	17	19

Fig.5.6 Array of series coefficients.

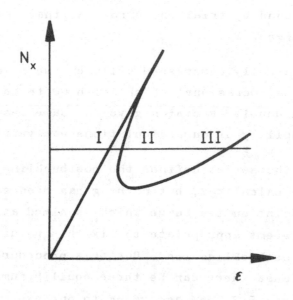

Fig.5.7 Three equilibrium states at one load.

The calculation is uniquely and reliably controlled, if one selects as a control value the coefficient of the dominant term of the series. This value is kept constant during the iteration. In Fig.5.6 the constant control value is marked by hatchings. In order to keep the number of unknowns equal to the number of equations, the postbuckling load N_x will be treated as an unknown parameter.

This kind of calculation is not only used for the calculation of the advanced postbuckling region, but also for the calculation of the initial postbuckling region by the perturbation method.

5.1.5 Solution of the compatibility condition

The solution procedure is outlined for the series expansion (27). The first step in solving the nonlinear differential equations consists of expressing the coefficient $b_{\tilde{j}\tilde{m}}$ in terms of the coefficients a_{jm} by means of the compatibility equation (18). As a preparation, the series (27a) for the deflection w is introduced in the nonlinear terms of the compatibility equation. As an example the first nonlinear term is treated

$$\frac{\partial^2 w}{\partial x^2} \cdot \frac{\partial^2 w}{\partial y^2} = \frac{1}{2}\left(\frac{\partial^2 w}{\partial x^2} \cdot \frac{\partial^2 \overline{w}}{\partial y^2} + \frac{\partial^2 \overline{w}}{\partial x^2} \cdot \frac{\partial^2 w}{\partial y^2}\right)$$

(33)
$$= \frac{1}{2} \sum_j \sum_m \sum_{\bar{j}} \sum_{\bar{m}} \cdot a_{jm} \cdot a_{\bar{j}\bar{m}} \cdot$$

$$\cdot \left(K_{jm\bar{j}\bar{m}} + K_{\bar{j}\bar{m}jm}\right) \cdot \cos\frac{m\pi x}{L} \cdot \cos\frac{\bar{m}\pi x}{L} \cdot \cos\frac{jny}{r} \cdot \cos\frac{\bar{j}ny}{r} \ ,$$

with

(33a)

$$K_{jm\bar{j}\bar{m}} = \left(\frac{m\pi}{l}\right)^2 \cdot \left(\frac{\bar{j}n}{r}\right)^2$$

$$K_{\bar{j}\bar{m}jm} = \left(\frac{\bar{m}\pi}{l}\right)^2 \cdot \left(\frac{jn}{r}\right)^2 .$$

The unknown coefficients a_{jm} are split up into a finitely high estimated value $(a_{jm})_o$ and an increment Δa_{jm} which is assumed to be so small that products of increments are negligible.

(34)
$$a_{jm} = (a_{jm})_0 + \Delta a_{jm} .$$

Introducing (34) into the first nonlinear term as written in (33) yields

$$\frac{\partial^2 w}{\partial x^2} \cdot \frac{\partial^2 w}{\partial y^2} = \frac{1}{2} \sum_j \sum_m \sum_{\bar{j}} \sum_{\bar{m}} \left[(a_{jm})_0 (a_{\bar{j}\bar{m}})_0 + (a_{jm})_0 \cdot \Delta a_{\bar{j}\bar{m}} + (a_{\bar{j}\bar{m}})_0 \cdot \Delta a_{jm} \right] \cdot$$

$$\cdot (K_{jm\bar{j}\bar{m}} + K_{\bar{j}\bar{m}jm}) \cos \frac{m\pi x}{l} \cdot \cos \frac{\bar{m}\pi x}{l} \cdot \cos \frac{ny}{r} \cdot \cos \frac{\bar{n}y}{r} .$$

This expression is symmetric in the pairs of indices (j,m) and $(\bar{j}, \bar{m})$. Therefore, these index pairs may be interchanged in the term $(a_{jm})_o \cdot \Delta a_{\bar{j}\bar{m}}$. This leads to

$$\frac{\partial^2 w}{\partial x^2} \cdot \frac{\partial^2 w}{\partial y^2} = \sum_j \sum_m \sum_{\bar{j}} \sum_{\bar{m}} \left[\frac{1}{2} (a_{jm})_0 (a_{\bar{j}\bar{m}})_0 + (a_{\bar{j}\bar{m}})_0 \cdot \Delta a_{jm} \right] \cdot$$

$$\cdot (K_{jm\bar{j}\bar{m}} + K_{\bar{j}\bar{m}jm}) \cdot \cos \frac{m\pi x}{l} \cdot \cos \frac{\bar{m}\pi x}{l} \cdot \cos \frac{jny}{r} \cdot \cos \frac{\bar{j}ny}{r} .$$

In this expression, the coefficients $(a_{jm})_o$ and $(a_{\bar{j}\bar{m}})_o$ are known values at each iteration step, the factors $K_{jm\bar{j}\bar{m}}$ and $K_{\bar{j}\bar{m}jm}$ (33a) depend on the buckle length and, hence, are also known, whereas Δa_{jm} are the unknown increments of the coefficients a_{jm}.

When the second nonlinear term is correspondingly modified and, together with the series (27) for the stress function ϕ and the radial displacement w, is substituted into the compatibility condition (18), one obtains

(35)

$$\sum_{\tilde{j}}^{\tilde{j}e} \sum_{\tilde{m}}^{\tilde{m}e} V_{\tilde{j}\tilde{m}} \left[(b_{\tilde{j}\tilde{m}})_0 + \Delta b_{\tilde{j}\tilde{m}} \right] \cos \frac{\tilde{m}\pi x}{l} \cdot \cos \frac{\tilde{j}ny}{r}$$

$$= \frac{1}{r} \sum_{j}^{je} \sum_{m}^{me} \left[(a_{jm})_0 + \Delta a_{jm} \right] \left(\frac{m\pi}{l} \right)^2 \cdot \cos \frac{m\pi x}{l} \cdot \cos \frac{jny}{r} -$$

$$- \sum_{j}^{je} \sum_{m}^{me} \sum_{\tilde{j}}^{\tilde{j}e} \sum_{\tilde{m}}^{\tilde{m}e} \left[\frac{1}{2} (a_{jm})_0 \cdot (a_{\tilde{j}\tilde{m}})_0 + (a_{\tilde{j}\tilde{m}})_0 \cdot \Delta a_{jm} \right] \cdot$$

$$\cdot \left[(K_{jm\tilde{j}\tilde{m}} + K_{\tilde{j}\tilde{m}jm}) \cdot \cos \frac{m\pi x}{l} \cdot \cos \frac{\tilde{m}\pi x}{l} \cdot \cos \frac{jny}{r} \cdot \cos \frac{\tilde{j}ny}{r} - \right.$$

$$\left. - S_{jm\tilde{j}\tilde{m}} \cdot \sin \frac{m\pi x}{l} \cdot \sin \frac{\tilde{m}\pi x}{l} \cdot \sin \frac{jny}{r} \cdot \sin \frac{\tilde{j}ny}{r} \right],$$

with

$$V_{\tilde{j}\tilde{m}} = A_{11} \left(\frac{\tilde{j}n}{r} \right)^4 + (A_{12} + 2A_{33}) \left(\frac{\tilde{m}\pi}{l} \right)^2 \cdot \left(\frac{\tilde{j}n}{r} \right)^2 + A_{22} \left(\frac{\tilde{m}\pi}{l} \right)^4$$

and

$$K_{jm\tilde{j}\tilde{m}} + K_{\tilde{j}\tilde{m}jm} = \left(\frac{m\pi}{l} \right)^2 \cdot \left(\frac{\tilde{j}n}{r} \right)^2 + \left(\frac{\tilde{m}\pi}{l} \right)^2 \cdot \left(\frac{jn}{r} \right)^2$$

$$S_{jm\tilde{j}\tilde{m}} = 2 \frac{m\pi}{l} \cdot \frac{\tilde{j}n}{r} \cdot \frac{\tilde{m}\pi}{l} \cdot \frac{jn}{r}.$$

It is noticable that the coefficients for the stress function

(36)
$$b_{\tilde{j}\tilde{m}} = (b_{\tilde{j}\tilde{m}})_0 + \Delta b_{\tilde{j}\tilde{m}}$$

as well as the unknown increments Δa_{jm} of the coefficients a_{jm} occur only linearly, and that the terms of the stress function which contain the uniformly distributed membrane forces

$$\overset{o}{N}_x \frac{y^2}{2} + p \cdot r \cdot \frac{x^2}{2}$$

and a_o defined in equation (28) have fallen out.

We split up equation (35) into an equation that only contains finitely large terms such as $(b_{\tilde{j}\tilde{m}})_o$ and $(a_{jm})_o$ and an equation that contains the increments $\Delta b_{\tilde{j}\tilde{m}}$ and $\Delta a_{\tilde{j}\tilde{m}}$. Both equations can be solved exactly by a comparison of coefficients, and one obtains as a result

(37)
$$\left(b_{\tilde{j}\tilde{m}}\right)_0 = F\left((a_{jm})_0 , (a_{\tilde{j}\tilde{m}})_0 \cdot (a_{jm})_0\right) ,$$
$$\Delta b_{\tilde{j}\tilde{m}} = f\left(\Delta a_{jm} , (a_{\tilde{j}\tilde{m}})_0 \cdot \Delta a_{jm}\right) .$$

If nothing is neglected in this comparison of coefficients, the range of indices $\tilde{j}$, $\tilde{m}$ will be twice the range of indices j , m owing to the products of trigonometric functions in the nonlinear terms

(38)
$$\tilde{j}e = 2je \quad \text{and} \quad \tilde{m}_e = 2m_e .$$

5.1.6 Solution of the equilibrium condition

The equilibrium condition (13) is solved following the Galerkin method. In doing so, one obtains as many equations

$$
\begin{aligned}
(39) \int_0^{2\pi r} \int_0^{l} &\left[D_{11} \cdot \frac{\partial^4 w}{\partial x^4} + 2\left(D_{12} + 2\,D_{33}\right) \frac{\partial^4 w}{\partial x^2 \partial y^2} + D_{22} \cdot \frac{\partial^4 w}{\partial y^4} - \frac{1}{r} \cdot \frac{\partial^2 \Phi}{\partial x^2} + p \right. \\
&\left. - \frac{\partial^2 \Phi}{\partial y^2} \cdot \frac{\partial^2 w}{\partial x^2} - \frac{\partial^2 \Phi}{\partial x^2} \cdot \frac{\partial^2 w}{\partial y^2} + 2 \frac{\partial^2 \Phi}{\partial x \partial y} \cdot \frac{\partial^2 w}{\partial x \partial y} \right] \cos \frac{m^* \pi x}{l} \cdot \cos \frac{j^* n y}{r}\, dx\, dy
\end{aligned}
$$

as unknown coefficients a_{jm} are existing in the series expansion (27a) for the radial deformation w .

If one substitutes the series (27) for the radial displacement w and the stress function ϕ with coefficients split up according to equation (34) and (36) into the equation (39), performs the integration, eliminates the coefficients $b_{\widetilde{jm}}$ and $\Delta b_{\widetilde{jm}}$ of the stress function by means of the relation (37) obtained from the compatibility condition and sets to zero all quantities which are small of second order, one obtains a linear equation system for the calculation of the unknown increments Δa_{jm} .

In chapter 5.1.3 it has been stated that the dominant term a_{jm} is used as a control value that prescribes the point on the postbuckling curve which is aimed at. Hence, the increment of this coefficient is zero. In order to complete the number of unknowns the postbuckling load $\overset{\circ}{N}_x$ is split up into an estimate $\overset{\circ}{N}_{xo}$ and a small increment $\Delta \overset{\circ}{N}_x$,

$$
(40) \qquad\qquad \overset{\circ}{N}_x = \overset{\circ}{N}_{xo} + \Delta \overset{\circ}{N}_x
$$

and $\Delta \overset{\circ}{N}_x$ is treated as an unknown quantity.

So a linear equation system has been built up, the unknowns of which are the increment $\Delta \overset{\circ}{N}_x$ of the postbuckling load $\overset{\circ}{N}_x$

and the increments Δa_{jm} of the coefficients a_{jm} . The
assumption that the increments $\Delta \overset{\circ}{N}_x$ and Δa_{jm} are small
is generally not fullfilled for the first estimates. Hence,
the equation system can be solved only by iteration.

In the program the following three input values control the
iteration:

- First, the number of series terms with which the
  calculation begins.

- Second, the increment with which the number of terms
  is increased, when the iteration for a given number
  of series terms has converged, i.e. when the incre-
  ment $\Delta \overset{\circ}{N}_x$ of the postbuckling $\overset{\circ}{N}_x$ has become negligibly small.

- Third, the maximum number of series terms. This upper
  limit should be so high that the addition of further
  terms would no longer influence the postbuckling load
  and should be so low that the demand for storage and
  computing time remains tolerable.

5.1.7 Shortening and potential energy

It is explained in section 3.2 that the equilibrium states
that occur in axially loaded cylinders during test, are
characterized by particularly small shortenings or particularly small potential energies.

For unpressurized cylinders we calculated the smallest end-
shortenings attained by a number of postbuckling patterns.
The mean shortening averaged over the distance between two
planes of symmetry follows from equations (17a) and (15a)
with (12) as

(41) $\quad \varepsilon(y) = \dfrac{u(y)}{l} = \dfrac{1}{l}\displaystyle\int_0^l \left[A_{11}\cdot\dfrac{\partial^2\Phi}{\partial y^2} + A_{12}\cdot\dfrac{\partial^2\Phi}{\partial x^2} - \dfrac{1}{2}\left(\dfrac{\partial w}{\partial x}\right)^2 \right] dx .$

The end-shortening given by equation (41) is formally a function of the circumferential coordinate y . Its mean value along the circumference is

(42)
$$\varepsilon = \dfrac{1}{2\pi r}\displaystyle\int_0^{2\pi r} \varepsilon(y)\, dy$$

$$= A_{11}\cdot \overset{\circ}{N}_x - \dfrac{1}{8}\displaystyle\sum_j\sum_m a_{jm}^2\left(\dfrac{m\pi}{l}\right)^2\cdot\beta ,$$

with

$$\beta = \quad 2 \quad \text{for} \quad j = 0$$
$$\beta = \quad 1 \quad \text{for all the other cases} .$$

The use of this mean value is justified since the cylinder section under consideration is bounded by planes of symmetry. Our attempt to show without averaging that the periodic terms in equation (41) cancel out, did not succeed.

For pressurized cylinders we calculate the potential energy

(43)
$$P = \dfrac{1}{2}\displaystyle\int_0^l\int_0^{2\pi r}\left(N_x\cdot\varepsilon_x + N_y\cdot\varepsilon_y + N_{xy}\cdot\gamma_{xy}\right.$$

$$\left. + M_x\cdot\dfrac{\partial^2 w}{\partial x^2} + M_y\cdot\dfrac{\partial^2 w}{\partial y^2} + M_{xy}\cdot 2\,\dfrac{\partial^2 w}{\partial x\partial y}\right) dx\, dy$$

$$+ \displaystyle\int_0^l\int_0^{2\pi r} p\cdot w\cdot dx\, dy .$$

From equation (43), with equations (9), (12), (15) and (27)
results

$$\frac{P}{\pi r l} = A_{11} \cdot \overset{\circ}{N}_x^2 + 2 A_{12} \cdot \overset{\circ}{N}_x \cdot pr + A_{22} (pr)^2$$

(44)
$$+ \frac{1}{4} \sum \sum b_{\widetilde{j}\widetilde{m}}^2 \cdot V_{\widetilde{j}\widetilde{m}} + \frac{1}{4} \sum \sum a_{jm}^2 \cdot W_{jm}$$

$$+ 2 p \left[a_0 + \frac{1}{4r} \cdot \sum (a_{j0} \cdot j n)^2 \right] ,$$

with the abbreviations

(45)
$$V_{\widetilde{j}\widetilde{m}} = \left[A_{11} \left(\frac{\widetilde{j}n}{r} \right)^2 + (2 A_{12} + A_{33}) \left(\frac{\widetilde{m}\pi}{l} \right)^2 \cdot \left(\frac{\widetilde{j}n}{r} \right)^2 + A_{22} \left(\frac{\widetilde{j}n}{r} \right)^4 \right] \cdot \mathcal{A}$$

$$W_{jm} = \left[D_{11} \left(\frac{m\pi}{l} \right)^4 + (2 D_{12} + 2 D_{33}) \left(\frac{m\pi}{l} \right)^2 \cdot \left(\frac{jn}{r} \right)^2 + D_{22} \left(\frac{jn}{r} \right)^4 \right] \cdot \mathcal{A}$$

and with

$$\mathcal{A} = 1 \quad \text{for} \quad \widetilde{j} \neq 0 \quad \text{and} \quad \widetilde{m} \neq 0 \quad \text{and}$$
$$\text{for} \quad j \neq 0 \quad \text{and} \quad m \neq 0$$

$$\mathcal{A} = 2 \quad \text{for} \quad \widetilde{j} = 0 \quad \text{or} \quad \widetilde{m} = 0 \quad \text{and}$$
$$\text{for} \quad j = 0 \quad \text{or} \quad m = 0 .$$

5.1.8 Execution of calculations

When performing the calculation it has proved useful to
begin the first postbuckling curve of a diagram in the
vicinity of the bifurcation point, where the amplitudes
of the postbuckling pattern are very small. That means:
to prescribe a very small control value a_{11} , to set all
unknown values $(a_{jm})_0$ to zero and to use the buckling
load as an estimate for the postbuckling load $\overset{\circ}{N}_{xo}$. The
iteration procedure generally is started with about 30 terms.

When the iteration is concluded, that means, when the postbuckling load $\overset{\circ}{N}_x$ has converged to a given accuracy the
number of terms is increased until the maximum number is
reached.

Thus the first point of the first postbuckling curve has
been found. For the next point, one alters only the control
value a_{11} ; the postbuckling load $\overset{\circ}{N}_x$ and the coefficients
a_{jm} of the calculated point are used as estimated values.
The calculation is performed with the maximum number of
terms from the beginning. If the steps, i.e. the increments
of the control value a_{11} are small enough, one finds the
next point within 2 iterations.

In proceeding to the next postbuckling curve, one alters
only the characteristic value of the next curve, for instance the circumferential wave number n or the cylinder
length 1 . The control value a_{11} and all estimated values
are retained.

If only the smallest postbuckling shortenings or the smallest postbuckling loads are sought, it is sufficient to calculate short pieces of the postbuckling curves.

As mentioned before, nonlinear differential equations may
have a great variety of solutions. Years of practical experience with the solution of Donnell's shell equations
taught us that physically meaningless results, e.g. tensile postbuckling loads, are obtained with iterative calculations only if the series expansion for the solution
contained so few coefficients that a physically reasonable
equilibrium state could not be obtained. Otherwise the
iteration either converged to a physically reasonable equilibrium state or did not converge at all.

5.2 Spherical shells

5.2.1 Notations

$A_{\tilde{n}}$ Coefficients of stress function,
         defined in equ. (72)

D   Bending stiffness, for isotropic shells

E   Young's modulus

F   Stress function, defined in equ. (47)

H, V   Horizontal and vertical components of stress
         resultants, defined in Fig.5.8

K_{ij} Stiffness coefficients   (i,j = 1,2)

M_φ Bending moment in meridional direction,
         defined in Fig.5.9

n, N   Integers

N_φ Membrane force in meridional direction,
         defined in Fig.5.9

P   Total potential energy, defined in
         equs. (51) and (66)

U   Strain energy, defined in equs. (52) and (54)

V   Potential of external forces, defined in
         equs. (57) and (60)

a   Radius of the spherical shell

b_n Coefficients of series for β_a ,
         defined in equ. (70)

t   Thickness of shell

p   External pressure

u, w   Horizontal and vertical displacements,
         defined in Fig.5.8

v   Volume displaced during deformation

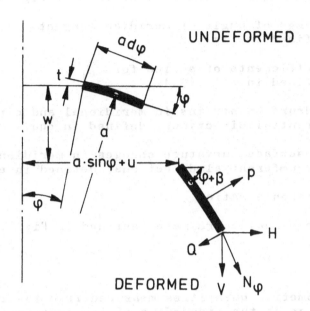

Fig.5.8   Dimensions, co-ordinates, displacements, and
forces acting at shell section.

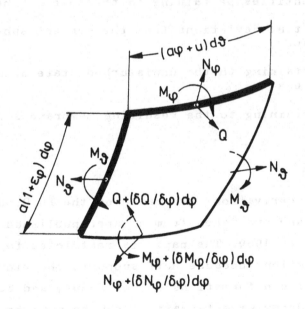

Fig.5.9   Stress resultants.

α Edge angle of the cap, defined in Fig.5.11

β Change of angle of meridian tangent, defined in Fig.5.8

δ_n Coefficients of series for β_i, defined in equ. (71)

$\varepsilon_\varphi, \varepsilon_\vartheta$ Midsurface strains in meridional and circumferential directions, defined in equ. (46)

$K_\varphi; K_\vartheta$ Midsurface curvature changes in meridional and circumferential directions, defined in equ. (46)

ν Poisson's ratio

φ Meridional coordinate, defined in Fig.5.8

Subscripts

a   Geometric quantities measured from the initial shape of the midsurface of the imperfect shell

b   Pertaining to the disturbance of the state of membrane stress in the shell, caused by buckling

e   Quantities pertaining to the edge of the shell

i   Initial deviations from the perfect spherical shape

m   Pertaining to the undisturbed state of membrane stress

$\bar{m}$ Pertaining to the resulting membrane stresses

5.2.2 Survey

The following derivations are based on the Ph.D. dissertation by Koga and are taken from a paper published by Koga and Hoff [13] in 1969. The paper is restricted to axisymmetric deformations because Hoff supports the view that non-axisymmetric deformations do not occur, and because axisymmetric decaying deformations can be more easily

calculated than a postbuckling pattern that consists of a
number of buckles extending regularly about the whole spherical surface.

The aim of Koga's investigations was the determination of
buckling loads of imperfect complete spheres buckling in an
axisymmetric pattern. Incidentally, complete and continuous
postbuckling curves were obtained in the computations. The
paper begins with a detailed deduction of the formulas.

In the analysis the complete spherical shell is divided into two parts:

- The first part is a shallow spherical cap in the center of which the axisymmetric initial imperfection
  exists and in which the dimple forms. It is assumed
  that the deformations are large in this part. Hence,
  the nonlinear theory of finite deflections is used
  for the analysis.

- The second part is the remaining portion of the sphere
  most of which deforms in simple contraction. The bending deformations of the remainder occur in an edge
  zone and are assumed to be small. Hence, the linear
  bending theory may be used for the analysis; the effect of the initial stress of the fundamental state
  is duly considered.

5.2.3 Basic equations

The basic equations governing axisymmetric large deformations but small strains of rotationally symmetric shells
were formulated by E. Reissner [14]. They have been simplified by Koga for shallow spherical shells with finite displacements and small rotations of the meridian. If there

exists a small initial imperfection characterized by the
initial angle β_i ("i" for initial) between the tangents
to the meridian of the imperfect and perfect shells, and
if the additional displacements and rotations measured
from the midsurface of the initially imperfect spherical
shell are designated by the subscript "a" (for additional),
the strain-displacement relations become

$$\varepsilon_{\varphi a} = \frac{d u_a}{a \cdot d\varphi} + \varphi \cdot \beta_a + \frac{1}{2} \beta_a^2 + \beta_i \, \beta_a \qquad \varepsilon_{\vartheta a} = \frac{u_a}{a \cdot \varphi} \,,$$

(46a-d)

$$\varkappa_{\varphi a} = -\frac{d \beta_a}{a \cdot d\varphi} \,, \qquad\qquad\qquad \varkappa_{\vartheta a} = -\frac{\beta_a}{a \cdot \varphi} \,,$$

where u_a is the horizontal component of displacement,
Fig.5.8 .

A stress function is introduced defined by the equations

(47a-b) $\qquad F = a \cdot \varphi \cdot N_\varphi \qquad\qquad \dfrac{d F}{d\varphi} = a \cdot N_\vartheta$.

Then the strains $\varepsilon_{\varphi a}$, $\varepsilon_{\vartheta a}$ and the deflection u_a can be
written in terms of F

$$\varepsilon_{\varphi a} = \frac{1}{E \cdot t \cdot a} \left(\frac{F}{\varphi} - \nu \cdot \frac{d F}{d\varphi} \right)$$

(48a-c) $\qquad\qquad \varepsilon_{\vartheta a} = \dfrac{1}{E \cdot t \cdot a} \left(\dfrac{d F}{d\varphi} - \dfrac{\nu \cdot F}{\varphi} \right)$

$$u_a = \frac{1}{E \cdot t} \left(\varphi \cdot \frac{d F}{d\varphi} - \nu \cdot F \right) \,,$$

where E , ν and t are Young's modulus. Poisson's ratio
and wall-thickness of the shell respectively.

5.2.4 Compatibility equation

The compatibility equation is formulated by expressing the combination of strains

(49)
$$\varepsilon_{\vartheta a} + \varphi \cdot \frac{d\varepsilon_{\vartheta a}}{d\varphi} - \varepsilon_{\varphi a}$$

by the displacements according to equation (46) on the one hand and by the membrane forces, equation (48), on the other hand, and then equating these expressions

(50)
$$\boxed{\frac{1}{E \cdot t \cdot a} \left(\frac{d^2 F}{d\varphi^2} + \frac{1}{\varphi} \cdot \frac{dF}{d\varphi} - \frac{1}{\varphi^2} \cdot F \right) = -\beta_a - \frac{1}{2\varphi} \cdot \beta_a^2 - \frac{1}{\varphi} \cdot \beta_a \beta_i}$$

The compatibility equation (50) is a nonlinear differential equation for the two dependent variables F and β_a .

5.2.5 Total potential energy

The total potential energy of the spherical cap can be separated into the following four parts:

$U_{\overline{m}}$ = Strain energy of the membrane forces

U_b = Strain energy of the bending moments

V_p = Potential energy of the external pressure acting upon the cap

V_e = Potential energy of the edge forces

(51)
$$P = U_{\overline{m}} + U_b + V_p + V_e .$$

The strain energy of the membrane forces is

(52)
$$U_{\overline{m}} = \pi \cdot a^2 \int_0^\alpha \left[N_\varphi \cdot \varepsilon_{\varphi a} + N_\vartheta \cdot \varepsilon_{\vartheta a} \right] \varphi \cdot d\varphi .$$

Substituting equations (48a-b) and equation (47) into equation (52) gives

$$(53) \qquad U_{\overline{m}} = \frac{\pi}{E \cdot t} \int_0^\alpha \left[\left(\frac{F}{\varphi}\right)^2 + \left(\frac{dF}{d\varphi}\right)^2 - 2\nu \cdot \frac{F}{\varphi} \cdot \frac{dF}{d\varphi} \right] \varphi \cdot d\varphi .$$

The strain energy of the bending moments is

$$(54) \qquad U_b = \pi \cdot a^2 \int_0^\alpha \left[M_\varphi \cdot \varkappa_{\varphi a} + M_{\vartheta} \cdot \varkappa_{\vartheta a} \right] \varphi \cdot d\varphi .$$

For the bending moments,

$$(55a\text{-}b) \qquad \begin{aligned} M_\varphi &= -D\left(\varkappa_\varphi + \nu \cdot \varkappa_\vartheta\right) \\ M_\vartheta &= -D\left(\varkappa_\vartheta + \nu \cdot \varkappa_\varphi\right) \end{aligned}$$

applies. Substituting equation (55a-b) with (46c-d) in equation (54), one obtains

$$(56) \qquad U_b = \pi \cdot a^2 \, D \int_0^\alpha \left[\left(\frac{d\beta_a}{a \cdot d\varphi}\right)^2 + \left(\frac{\beta_a}{a \cdot \varphi}\right)^2 + 2\nu \frac{\beta_a}{a \cdot \varphi} \cdot \frac{d\beta_a}{a \cdot d\varphi} \right] \varphi \cdot d\varphi .$$

The potential of the external pressure acting upon the cap is given in the shallow shell approximation by

$$(57) \qquad V_P = -2\pi a^2 p \int_0^\alpha w_a \cdot \varphi \cdot d\varphi ,$$

where w_a is the vertical component of the displacements measured from the midsurface of the initially imperfect shell and p is the external pressure. Integrating equation (57) by parts, one obtains

$$V_P = -\pi \left(a\alpha\right)^2 \cdot p \cdot w_a\left(\alpha\right) + \pi \cdot a^2 \cdot p \int_0^\alpha \frac{dw_a}{d\varphi} \cdot \varphi^2 \cdot d\varphi .$$

Without loss of generality the vertical displacement of the
edge of the cap can be set

$$w_a \left(\alpha \right) = 0 \ .$$

This means physically that the cross section of the sphere
in which the cap and the remainder adjoin, is not shifted
in the vertical direction during the buckling process. Thus
one has

(58)
$$V_p = \pi \cdot a^3 \cdot p \int\limits_0^\alpha \varphi^2 \cdot \beta_a \cdot d\varphi \ ,$$

where the relation

(59)
$$\beta_a = \frac{dw_a}{a \cdot d\varphi}$$

has been used.

In formulating the expression for the potential energy of
the edge forces it will be distinguished between

- the membrane forces in the undisturbed spherical
  shell and the corresponding deformations; these will
  be called "membrane" in the following and marked by
  the index "m" .

- the edge forces and moments due to the bending of
  the shell, caused by buckling, and the corresponding
  deformations; these will be called "bending" in the
  following and marked by the index "b" .

When calculating the potential energy of the edge forces,
those parts not containing the variables F and β_a are
neglected, as they have no influence on the subsequent

minimization of the potential energy. The remaining potential consists of three parts:

- the "bending" force H_{be}
  acting on the "bending" deflection u_{be}

- the "membrane" force H_{me}
  acting on the "bending" deflection u_{be}

  the "bending" moment $M_{\varphi e}$
  acting on the angle of rotation β_{ae} .

Utilizing the sign convention of Fig.5.10 one obtains the
potential energy of the edge forces

(60)
$$V_r = 2\pi \alpha a \left[-\frac{1}{2} H_{be} \cdot u_{be} - H_{me} \cdot u_{be} + \frac{1}{2} M_{\varphi e} \beta_{ae} \right]$$
$$= \pi \alpha a \left[-\left(H_{be} + 2 H_{me} \right) u_{be} + M_{\varphi e} \beta_{ae} \right] .$$

Within the scope of the shallow shell theory, the "membrane"
force is

(61)
$$H_{me} = -\frac{pa}{2} ,$$

the "bending" force is

(62)
$$H_{be} = N_{\varphi e} - H_{me} = N_{\varphi e} + \frac{pa}{2}$$

and the "bending" deflection is

(63)
$$U_{be} = U_{be} - U_{me} = U_{ae} + \frac{pa\,(1-\nu)}{2\,E\cdot t} \cdot \alpha \cdot a .$$

Substituting the equations (61), (62) and (63) into equation (60) yields

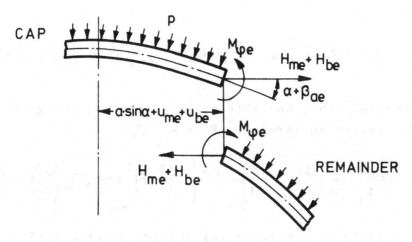

Fig.5.10 Sign convention for edge loads and edge
          displacements.

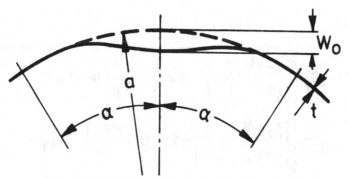

Fig.5.11 Imperfection shape.

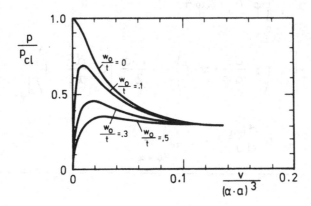

Fig.5.12 Calculated pressure-volume curves.

(64) $V_r = \pi \alpha\, a \left[-\left(N_{\varphi e} - \dfrac{pa}{2} \right) \left(u_{ae} + \dfrac{pa\,(1-\nu)}{2 \cdot E \cdot t}\, \alpha\, a \right) + M_{\varphi e}\, \beta_{ae} \right].$

From Equation (64), one obtains the potential energy of
the edge forces in terms of   F   and β_a :

(65) $V_n = \dfrac{\pi \alpha \alpha}{E \cdot t} \left[-\left(\dfrac{F}{a} \cdot \dfrac{dF}{d\varphi} - \nu \dfrac{F^2}{\alpha a} \right) + \dfrac{pa}{2} \left(\alpha \cdot \dfrac{dF}{d\varphi} - F \right) \right]_{\varphi = \alpha} + \pi \alpha\, D \beta_{ae} \left(\dfrac{d\beta_a}{d\varphi} + \nu \dfrac{\beta_{ae}}{\alpha} \right)_{\varphi = \alpha}.$

The four parts of the potential energy, equations (53),
(56), (58) and (65), established before, are now added to
yield

(66)

$$P = \frac{\pi}{E \cdot t} \int_0^\alpha \left[\left(\frac{dF}{d\varphi} \right)^2 + \left(\frac{F}{\varphi} \right)^2 - 2\nu\, \frac{F}{\varphi} \cdot \frac{dF}{d\varphi} \right] \varphi \cdot d\varphi +$$

$$+ \pi \cdot D \int_0^\alpha \left[\left(\frac{d\beta_a}{d\varphi} \right)^2 + \left(\frac{\beta_a}{\varphi} \right)^2 + 2\nu \cdot \beta_a \cdot \frac{d\beta_a}{d\varphi} \right] \varphi \cdot d\varphi +$$

$$+ \pi \cdot a^3 \cdot p \int_0^\alpha \varphi^2\, \beta_a \cdot d\varphi +$$

$$+ \frac{\pi}{E \cdot t} \left[-F \left(\alpha \cdot \frac{dF}{d\varphi} - \nu \cdot F \right) + \frac{pa^2 \cdot \alpha}{2} \left(\alpha \cdot \frac{dF}{d\varphi} - F \right) \right]_{\varphi = \alpha} +$$

$$+ \pi D \beta_{ae} \left(\alpha\, \frac{d\beta_a}{d\varphi} + \nu \cdot \beta_{ae} \right)_{\varphi = \alpha}.$$

5.2.6 Matching conditions

The assumption of small deformations for the remainder allows to make use of the well established solutions of the linear equations governing the deformations of the spherical shell. H_{be} and $M_{\varphi e}$ are related linearly to u_e and β_e. The relations are written in the form

(67)
$$H_{be} = K_{11} \cdot u_{be} + K_{12} \cdot \beta_{ae}$$
$$M_{\varphi e} = K_{21} \cdot u_{be} + K_{22} \cdot \beta_{ae} \,,$$

where K_{ij} $(i,j = 1,2)$ are stiffness coefficients for the remainder, which are presented in $[13,14,15]$.

After substituting equations (62), (63) and (55a) into equation (67), one obtains the matching conditions as functions of F and β_a

(68)
$$\left(\frac{F}{a \cdot \varphi}\right)_{\varphi = \alpha} + \frac{pa}{2} = \frac{K_{11}}{E \cdot t} \cdot \left(\alpha \cdot \frac{dF}{d\varphi} - \nu \cdot F + \frac{1-\nu}{2} \cdot p\alpha a^2\right)_{\varphi = \alpha} + K_{12} \, \beta_{ae}$$

(69)
$$\frac{D}{a} \left(\frac{d\beta_a}{d\varphi} + \nu \frac{\beta_a}{\alpha}\right)_{\varphi = \alpha} = \frac{K_{21}}{E \cdot t} \cdot \left(\alpha \frac{dF}{d\varphi} - \nu \cdot F + \frac{1-\nu}{2} \cdot p\alpha a^2\right)_{\varphi = \alpha} + K_{22} \cdot \beta_{ae} \,.$$

5.2.7 Method of solution

The nonlinear problem can only be solved approximately. The rotation β_a is assumed in the form of a finite polynomial in φ with undetermined coefficients b_n .

(70)
$$\beta_a = \sum_{n=1}^{N} b_n \cdot \varphi^n \,,$$

with $n = 1, 3, 5 \ldots$.

272

It contains only odd powers of φ , because the deformations are axisymmetric.

The initial imperfection is also assumed to be given in the form of a finite polynomial

$$(71) \qquad \beta_i = \sum_{n=1}^{N} \delta_n \cdot \varphi^n ,$$

where the δ_n are prescribed for a particular imperfection shape. In practice, some of the δ_n will be chosen to be zero, but having the same upper limit N in the sums of equation (70) and (71) is convenient when programming for the computer.

β_a and β_i as assumed in the form of equations (70) and (71) are substituted into the right hand member of the compatibility equation (50). Then equation (50) can be easily solved. The result can be written in the form

$$(72) \qquad F = \sum_{\tilde{n}=1}^{2N+1} A_{\tilde{n}} \cdot \varphi^{\tilde{n}} .$$

For $\tilde{n} > 1$ one obtains

$$(73) \qquad A_{\tilde{n}} = f_{\tilde{n}}(b_n, \delta_n) .$$

$A_{\tilde{1}}$ is still undefined. It will be determined with the aid of one of the matching conditions, say equation (68)

$$(74) \qquad A_{\tilde{1}} = f_{\tilde{1}}(b_n, \delta_n) .$$

Substituting the series (70) and (72) into equation (66) one obtains the total potential energy written in terms of b_n and $A_{\tilde{n}}$.

$$(75) \qquad P = f_{n \, \tilde{n}} \left(b_n , A_{\tilde{n}} \right) .$$

For given values of p , α and β_i the total potential energy P must be a minimum with respect to all b_n in the presence of the second matching condition (the first one was used to determine $A_{\tilde{1}}$) equation (69). The minimization yields a set of nonlinear algebraic equations in b_n , which is solved with the aid of the Newton iterative procedure.

Once the b_n are obtained, the $A_{\tilde{n}}$ are calculated according to equations (73) and (74). Then the basic quantities β_α , u_a , w_a , H and M_φ can be calculated for the cap.

The volume displacement during deformation is given by

$$(76) \qquad \frac{V}{(\alpha \cdot a)^3} = -2 \int_0^\alpha \beta_\alpha \cdot \varphi^2 \cdot d\varphi .$$

5.2.8 Numerical results

Comparative calculations have shown that an initial imperfection according to Fig.5.11 results in the lowest buckling load if the dimensionless opening angle is

$$(77) \qquad \alpha = 4 \sqrt{\frac{t}{a}} \cdot \frac{1}{\sqrt[4]{12 \, (1 - \nu^2)}} .$$

The evaluation of this formula in the case of a wall-thickness ratio $a/t = 1000$ yields a real opening angle of

$$2\alpha = 8° .$$

274

Fig.5.12 shows load-volume change curves which were calculated according to the previously described method. It
is remarkable that the buckling load is reduced to 35 %
of the buckling load of the perfect sphere with an initial
imperfection the amplitude of which is half as large as
the wall-thickness. According to this postbuckling diagram,
a further reduction of the buckling load by initial imperfections is scarcely possible, because the lowest calculated
postbuckling load amounts to 30 % of the buckling load of
the pefect sphere.

The diagram shows a dimensionless representation and is
therefore valid for all isotropic spherical shells. The
authors [13] had assumed the amplitude of the initial imperfections such that the theoretical buckling loads lie within the scatter of the experimental ones. Having attained
this result they were satisfied. However, the theoretical
and experimental imperfection amplitudes, postbuckling loads
and postbuckling patterns have not been compared.

Literature

[1]   DONNELL, L.H.   A new Theory for the Buckling of
                        Thin Cylinders under Axial Com-
                        pression and Bending.
                        Trans. ASME Vol.56 (1934) pp.795-806

[2]   FLüGGE, W.   Statik und Dynamik der Schalen.
                        Springer-Verlag, Berlin-Göttingen-
                        Heidelberg (1957), 286 S.
                        zweite neubearbeitete Auflage

[3]   THIELEMANN, W.F. New Development in the Nonlinear
                        Theories of the Buckling of Thin
                        Cylindrical Shells.
                        "Aeronautics and Astronautics",
                        Pergamon Press (1960) pp.76-121

[4] APPEL, H.   Axialsymmetrische Verformungen von
     GEIER, B.   exzentrisch versteiften orthotropen
                      Kreiszylinderschalen.
                      DLR FB 67-82 (1967), 68 S.

[5] GEIER, B.   Das Beulverhalten versteifter
                      Zylinderschalen.
                      Teil 1: Differentialgleichungen.
                      Z.Flugwiss. Bd.14 (1966), S.306-490

[6] MARGUERRE, K.   Zur Theorie der gekrümmten Platte
                      großer Formänderung.
                      Jahrb. d. deutschen Luftfahrtforsch.
                      (1939), S.I 413-I 418

[7] KOITER, W.T.   On the Nonlinear Theory of Thin
                      Elastic Shells.
                      Lab. of Eng. Mech., Dept. of Mech.
                      Engng., Technol. Univ., Delft,
                      Netherlands, Rep. No. 310

[8] KOITER, W.T.   On the Stability Elastic Equilibrium.
                      NASA TT F-10, 833

[9] BUDIANSKY, B.   Dynamic Buckling of Elastic Struc-
                      tures: Criteria and Estimates.
                      NASA-CR 66072 (1965), 49 pp.

[10] BUDIANSKY, B.   Dynamic Buckling of Imperfection
     HUTCHINSON, J.W. Sensitive Structures.
                      Harvard University, Cambridge, Mass.,
                      Eng. & App. Phys. Div. TR-18, June
                      (1964), 40 pp.

[11] ESSLINGER, M.   Nachbeulrechnung für einen unendlich
                      langen isotropen Kreiszylinder.
                      DLR-FB 72-37 (1972). 42 S.

[12] ESSLINGER, M.   Ein Verfahren zur theoretischen
                      Untersuchung des Beul- und Nachbeul-
                      verhaltens dünnwandiger Kreiszylinder
                      mit eingespannten Rändern.
                      DLR-FB 68-70 (1968), 127 S.

276

[13] KOGA, T.   The Axisymmetric Buckling of Ini-
     HOFF, N.J.   tially Imperfect Complete Spherical
                     Shells.

                     Int. J. Solids and Struct. Vol.5
                     (1969), pp.679-697

[14] REISSNER, E.   On Axisymmetrical Deformations of
                     Thin Shells of Revolution.

                     Proc. Symp. Appl. Math., Vol.III
                     (1950), pp.27-52, McGraw-Hill,
                     New York

[15] ESSLINGER, M.   Statische Berechnung von Kesselböden.

                     Springer-Verlag (1952), Dissertation

Contents

Printed in the United States
By Bookmasters